A Beginner's Guide to Immortality

ALSO BY CLIFFORD A. PICKOVER

A Beginner's Guide to Immortality

Extraordinary People, Alien Brains, and Quantum Resurrection

Clifford A. Pickover

Thunder's Mouth Press
New York

A BEGINNER'S GUIDE TO IMMORTALITY:
Extraordinary People, Alien Brains, and Quantum Resurrection

Copyright © 2007 by Clifford Pickover

Published by
Thunder's Mouth Press
An Imprint of Avalon Publishing Group, Inc.
245 West 17th Street, 11th Floor
New York, NY 10011

AVALON

LEGO® is a trademark of the LEGO® Danish toy company.

Library of Congress Cataloging-in-Publication Data is available.

ISBN-10: 1-56025-984-1
ISBN-13: 978-1-56025-984-8

9 8 7 6 5 4 3 2 1

Book design by Bettina Wilhelm

Printed in the United States of America
Distributed by Publishers Group West

This book is dedicated to all those
who do not have a book
dedicated to them.

This book is also dedicated to the cast
and filmmakers of *The Brain from Planet Arous*.
Readers are urged to purchase this movie
and to watch it while reading this book
and contemplating pop culture.

Every great work of art has two faces:
one toward its own time
and one toward the future, toward eternity.
—Daniel Barenboim, *Parallels and Paradoxes*

Seeing a chameleon in your dream
represents your ability to adapt to any situation.
You are versatile and are well-rounded.
Alternatively, you feel you
are being overlooked.
—*Dream Dictionary*[1]

That so few now dare to be eccentric,
marks the chief danger of our time.
—John Stuart Mill, *On Liberty*

I would not live forever, because we should not live forever,
because if we were supposed to live forever,
then we would live forever,
but we cannot live forever,
which is why I would not live forever.
—Miss Alabama in the 1994 Miss USA Contest,
answering the question, "If you could live forever, would you?"

Beauty is eternity gazing at itself in a mirror.
—Kahlil Gibran, *The Prophet*

CONTENTS

✣ ✤

ACKNOWLEDGMENTS

I thank Teja Krašek, Mark Nandor, Graham Cleverley, Mark Ganson, Dennis Gordon, Ray Ben Erskins, Jennifer Franklin Elrod, John Oakes, William Strachan, and members of the Clifford Pickover Think Tank (http://groups.yahoo.com/group/CliffordPickover/) for useful comments and suggestions.

Reference sources are listed at the end of the book. Several books were of particular value to me, including Michael Shermer's *Science Friction*, David Jay Brown's *Conversations at the Edge of the Apocalypse*, Stephen Mitchell's *Gilgamesh*, Glenn Yeffeth's *Taking the Red Pill*, A. J. Jacob's *The Know-It-All*, and Richard Metzger's *Book of Lies*. I describe some of the ideas of high-tech philosophers Nick Bostrom, Ray Kurzweil, and Robin Hanson in Chapter 4. For further reading on the fine line between genius and madness, see my book *Strange Brains and Genius: The Secret Lives of Eccentric Scientists and Madmen*. Several sections of the current book discuss powerful, illegal drugs, and I do not advocate or suggest that readers break any laws or try these drugs.

There are known knowns. There are things we know we know. We also know there are known unknowns. That is to say, we know there are some things we do not know. But there are also unknown unknowns, the ones we don't know we don't know.

—Secretary of Defense Donald Rumsfeld,
February 12, 2002, news briefing

Introduction

Life is a narrow vale between the cold and barren peaks of two eternities. We strive in vain to look beyond the heights. We cry aloud, and the only answer is the echo of our wailing cry. From the voiceless lips of the unreplying dead there comes no word; but in the night of death hope sees a star and listening love can hear the rustle of a wing.

—Robert G. Ingersoll, *The Ghosts and Other Lectures*

Chameleons. Such exceptional creatures. The way they change color. Red. Yellow. Lime. Pink. Lavender. And did you know they are very fond of music?

—Truman Capote, *Music for Chameleons*

A Celebration of Unusual Lives

After you die, will the world remember anything you did? Most of us rarely leave marks, except on our immediate family or a few friends. We'll never have our lives illuminated in a *New York Times* obituary or discussed by a TV news anchorperson.[1] Even your immediate family will know nothing of you within four generations. Your great-grandchildren may carry some vestigial memory of you, but that will fade like a burning ember when they die—and you will be extinguished and forgotten.

Even writers, like myself, don't have much of a chance. In fact, most best-selling books are destined to fade quickly. Consider, for example, how few of these hardcover bestsellers from the year 1950 are still in print or even remembered today.[2]

1. *The Cardinal*, Henry Morton Robinson
2. *Joy Street*, Frances Parkinson Keyes

3. *Across the River and into the Trees*, Ernest Hemingway
4. *The Wall*, John Hersey
5. *Star Money*, Kathleen Winsor
6. *The Parasites*, Daphne du Maurier
7. *Look Younger, Live Longer*, Gaylord Hauser
8. *Kon-Tiki*, Thor Heyerdahl
9. *Mr. Jones, Meet the Master*, Peter Marshall
10. *Campus Zoo*, Clare Barnes Jr.

Despite pop culture's tendency to foster the rapid extinction of books, ideas, and people, *A Beginner's Guide to Immortality* highlights unusual thinkers who punched through our ordinary cultural norms while becoming successful in their own niches. Here, we celebrate these extraordinary people and their curious ideas. Whether or not you recognize many of the names, these individuals dramatically changed the lives of those with whom they came in contact, and many have had a lasting impact on the world. In this book, I'll avoid the typical celebrities and influential Nobel laureates and instead focus on counterculture and more "right-brained" thinkers. Through these people, we can better explore life's astonishing richness and glimpse the diversity of human imagination.

Some of the "geniuses" in this book seem to have had peculiarities of one sort or another, or as Turin University professor Cesare Lomobroso once said, "Genius is often associated with anomalies in that organ which is the source of its glory." Their works and ideas often bear a personal mark, and a striving to tear apart traditional thinking or to make an impact—whether it was for their brilliant writing, for making the world's best chopped liver, or creating shocking new musical forms. Almost all of the people in this book had an irreverence toward authority and a self-sufficiency and independence. They were passionate about their work. Many of the trendsetters experienced social and professional resistance to their ideas. Most blazed a trail. I call these individuals "chameleons."

A Symphony for Chameleons

Chameleons are lizards famous for their ability to alter their skins to display an amazing variety of colors, ranging from blacks and browns to green, blue, yellow, red, or white. These odd creatures have startling eyes that can be moved independently of each other, allowing the lizards to survey the world with nearly 360-degree vision. Chameleons are the only lizards with zygodactyle feet, or pincers, that help them climb tall trees.

The people in this book are chameleons because they looked in many directions, were constantly changing, and usually had numerous interests. For example, musician John Cage was an avid mycologist and mushroom collector, and rocket-scientist Jack Parsons explored the farthest reaches of the occult. Psychologist William James was equally brilliant—teaching physiology, psychology, philosophy, and religion. Chameleon people reveal their creativity and humor in many ways. Many were part of countercultures, in opposition to mainstream society—iconoclasts driven by Promethean impulses to create. They tolerated ambiguity, were open to a cornucopia of experiences and areas of knowledge, were attracted to novelty, and wanted to influence the world. A typical chameleon would have the creativity and lateral thinking to answer this question if you handed it to them on a slightly soiled card:[3]

> Without using a pencil, how would you make this
> Roman numeral equation true?
> **XI + I = X**

When researching the chameleonic people for inclusion in this book, I was startled to find certain trends emerging. Many turned out to be homosexual, for example Truman Capote and John Cage, or appeared to have other nontraditional sexual lifestyles. A number of chameleons—including Truman Capote, John Cage, Jack Parsons, and the Wachowski Brothers—never completed college.

Going beyond the intriguing individuals, many of the *concepts* in the book are chameleonic, grabbing ideas from many fields such as mathematics, philosophy, zoology, and entertainment. We'll tackle quantum resurrection, the religious implications of mosquito evolution, simulated Matrix realities, the brain's own marijuana, and the mathematics of the apocalypse. If each area of human knowledge is likened to a spider web that glimmers in the sunlight, then these special topics come with unexpected connecting strands that unite the webs in a vast, sparkling fabric.

Pop Culture

Contrary to popular myth, when the chameleon lizard changes skin color, the morphing is not for the purpose of camouflage. The chameleon is not trying to fade away or blend into its surroundings. Rather, the remarkable colors are a reflection of the lizard's mood, temperature, and health. Chameleons use colors to communicate with others and express attitudes such as the willingness to mate or their determination to fight.

Similarly, the chameleon people in this book never blend with their environment, but rather show their flamboyant colors when promoting their ideas and at their creative heights. The zygodactyle chameleon grasps and does not let go. The chameleon people seize many ideas tightly and persuasively. They are all lateral thinkers—reasoning in directions not naturally pointed to by society or by the discipline in which they work. I also use the term "lateral thinking" in an extended way to indicate action motivated by serendipitous results, and the deliberate drift of thinking in new directions to discover what can be learned.

I sometimes aspire to being one of the chameleons. Each day, as I survey the world, I look in all directions. In this book, I often have one mental eye on a person while the other is considering related quirky facts. You'll find these digressions throughout this book.

The book is also about American popular culture. I love zany science-fiction movies from the 1950s, the recent history of ice cream empires and "Jewish" chopped liver, and beatniks who changed society by splashing new ideas onto the canvas of culture.

A New Age for America

If you are anything like me, your parents told you to get good grades, graduate from college, and pursue a career that offered a respectable living. Logical and analytical minds have always been at a premium. Careers as a physician, lawyer, engineer, or scientist were certainly good options in *my* home. However, a revolution in America is taking place in the early twenty-first century. According to writer Daniel H. Pink, we already have sufficient numbers of the linear, logical thinkers—including computer programmers and accountants whose tasks will be easily fulfilled by armies of Asian knowledge workers who increasingly contribute to American society. Pink believes:

> The future no longer belongs to people who can reason with computer-like logic, speed, and precision. It belongs to a different kind of person—[individuals with abilities like] artistry, empathy, seeing the big pictures, and pursuing the transcendent. . . . We're progressing to a society of creators and empathizers, pattern recognizers, and meaning makers.[4]

Pink's view is certainly controversial, but he is one of many who believes we are leaving the Information Age and entering the Conceptual Age. The most important people, and certainly the most interesting, will be those who create inventions that change our ways of life and break new ground. But more importantly, the hottest individuals will be those who are good at recognizing patterns in culture and belief, those who try to understand the forest and not just the trees. These pattern recognizers also help *others* become creative and dream daring dreams.

According to Professor Richard Florida of Carnegie Mellon

University, America and Europe are seeing a startling rise of what he calls the "Creative Class," people who are paid principally to do creative work for a living. Members of this class include scientists, engineers, artists, musicians, designers, and knowledge-based professionals. In 1900, fewer than 10 percent of American workers were doing creative work. Many people spent their daily grind in factories or on farms. However, by 2000, nearly a third of the workforce was part of the Creative Class. Florida writes in *The Rise of the Creative Class* that creative work accounts for half of all wage and salary income in the United States, over $1.7 trillion.

Come with me for a walk down Fifth Avenue in New York City. Do you want to make a lot of money? Many American entrepreneurs look beyond logic and consider worlds beyond our ordinary reality. Consider yoga as one example. Americans spend over $20 billion a year on yoga products such as $400 leather-trimmed yoga mats created by New York designer Marc Jacobs.[5] For no utilitarian reason, people still burn incense, light candles, and burn wood. Americans spend over $25 billion a year to enhance the lawns of their suburban homes. TV shows reflect society's interest in the offbeat, spooky, and transcendent. *Medium*, a TV show featuring a woman psychic, debuted in 2005 with over 16 million viewers. The popular show *The X-Files* featured the paranormal for seven seasons and led to a big-budget movie and loads of franchise merchandise. *Point Pleasant* featured a young woman in a New Jersey resort town who turns out to be the daughter of Satan. *Charmed* revolved around witches who battle their inner demons while tackling supernatural threats.

This interest in transcendent entertainment is clearly not limited to America. For example, consider the recent wave of spooky films that originated in Asia—*The Eye, The Ring, The Grudge, Dark Water, Pulse,* and other movies by the Pang Brothers, twin-brother screenwriters and film directors born in Hong Kong. According to John Hodgman, contributing writer to the *New York Times Magazine*, these films all suggest a growing appetite for "the supernatural reasserting itself within the rational world, often by the very technology that had abolished it (be it VHS tapes, modern medicine or, as in *Pulse*, the Internet. . . .)."[6]

Spiritually focused American TV shows are ubiquitous, including such past hits as *Touched by an Angel, Highway to Heaven, Revelations*, and *Joan of Arcadia*. As old shows fade away, new ones leap to take their places. The 2005 TV season led with such curiosities as *Supernatural* (ghosts, spirits, haunted woods), *Invasion* (mysterious people, strange lights in the swamp), *Night Stalker* (things that go bump in the night), *Surface* (mysterious underwater happenings), *Threshold* (aliens invade our dreams), and *Ghost Whisperer* (a woman senses earthbound spirits that need psychological closure before entering heaven).

Some of these 2005 shows were quickly canceled and sent straight to an infinite afterlife in DVD heaven. In 2006 and 2007, we saw a new crop of eerie TV movies or miniseries such as *The House Next Door* (successful suburban artist versus evil sprits) and *The Lost Room* (a motel-room key unlocks the door to a weird parallel world). Many of the themes of shows like *Charmed* were ancient, hearkening back to times when more people believed in miracles and fairies in the woods, felt a connection to nature, or used fire for sustenance.

Millions of Americans engage their atavistic impulses by investing in fireplaces for their homes or watching a televised "Yule log" burning on Christmas morning. Consider this amazing bit of Christmas trivia—TV station WPIX in New York City broadcasts a four-hour movie of a log blazing in a fireplace, which wins its time slot in the Nielsen ratings each year. The log never seems to be consumed because the movie is really just a repeating seven-minute tape loop. The original Yule log, first shown in 1966, was a black-and-white 17-second loop.[7] Why seek flames? Fire is a symbol of humanity's connection to nature—and the pseudorandom patterns of fire alter our brain waves, producing alpha waves and pleasure-producing neurochemicals.

The Butterfly Effect

I recently sat by my own fireplace as I finished watching *The Butterfly Effect*, a supernatural thriller in which psychology student Evan Treborn discovers that he can revisit his past and alter distressing events, hoping to improve their outcomes. However, his experiments have dreadful, personal consequences. Throughout the movie, Treborn surfs back and forth in time, witnessing variants of his life along multiple timelines. He discovers that fewer than ten words spoken in childhood can alter his life and, through the decades, perhaps the entire planet.

Chaos theory teaches us that our slightest actions can cause reality to change in profound ways. It's the "butterfly effect" that says the flapping of the wings of a monarch butterfly in New Jersey can change the weather in Iran a few weeks later, causing the downing of a power line, the fall of a repressive regime, and the blossoming of peace. Throughout this book, you'll also see butterflies in several contexts as symbols of sensitivity, transformation, and beauty.

I see this butterfly effect in the evolution of my own life, and perhaps you see it in yours. How did you get your first job? How did you choose your career or mate? It's obvious that history is contingent on the tiniest of forces. For example, imagine what might have happened if Cleopatra, Marie Antoinette, or Mata Hari had an ugly but benign skin growth on the tips of their noses. The entire cascade of historical events would be different. A mutation of a single skin cell—caused by the random exposure to sunlight—will change the universe. If everyone in the world about to have sex today delayed by a few seconds, all the resultant children conceived today would be *different* people, because a different sperm would penetrate the egg. *A one-second worldwide sneeze today would cause the conception of 720,000 new and different people on this day.*[8]

Even your own seemingly insignificant actions shape reality. A smile on a subway, a post on a Web bulletin board, or turning right instead of left alters the fabric of peoples' lives in unpredictable ways.

The chameleons in this book certainly influenced the world and touched countless lives.

Two poignant examples of the butterfly effect can be seen in World War II. The atomic bombing of Nagasaki, Japan, on August 9th, 1945, occurred only because the clouds over the primary target of Kokura were just a bit too thick, and the pilot had trouble finding the city's center. Over 80,000 were killed in Nagasaki and not Kokura because of a few clouds. Hitler survived an assassination attempt by German officer Ernst Stauffenberg because another officer nudged a briefcase bomb a few feet to the side of an oak table to keep things tidy. One of the table's two heavy supports shielded Hitler from the blast. Hitler survived, and millions died because of a subtle movement and a single piece of oak.

A more recent cultural example of the butterfly effect is exemplified by Norma McCorvey who, through a seemingly random incident in the 1970s, caused the massive crime drop of the 1990s in the US. University of Chicago economist Steven D. Levitt writes:

> Like the proverbial butterfly that flaps its wings on one continent and eventually causes a hurricane on another, Norma McCorvey dramatically altered the course of events without intending to. All she had wanted was an abortion. She was a poor, uneducated, unskilled, alcoholic, drug-using twenty-one-year-old woman who had already given up two children for adoption and now, in 1970, found herself pregnant again. But in Texas, as in all but a few states at that time, abortion was illegal.[9]

Several powerful people adopted McCorvey's cause and made her the lead plaintiff in a class action lawsuit seeking to legalize abortion. Her legal case eventually went to the US Supreme Court. In order to protect her privacy, her name was changed to Jane Roe, and in 1973, the court ruled in her favor, allowing legalized abortion in America.

So how did Roe's desire for an abortion cause the "greatest crime drop in recorded history"? Levitt points out that the millions of women most likely to have an abortion after *Roe v. Wade* were poor, unmarried, and teenage mothers for whom illegal abortions had been too risky and too expensive. And these mothers were the ones much more likely than average to produce criminals. But because of Roe, these potential thieves, drug dealers, murderers, and rapists were never born.[10] Before Roe, middle- and upper-class women could have relatively safe, illegal abortions. After Roe, any woman could have an abortion, safely, and for a price that a poor person could afford.

Truman Capote, John Cage, Jack Parsons, and Beyond

I'm a voracious reader and keep a diary of intriguing quotations that come across my line of sight each day. At the conclusion of each chapter of this book are quotations that relate to a book topic. These quotations continue in the appendix of the book titled "Cathedrals of the Mind," and they serve to kick-start additional lateral thinking. I welcome your feedback and look forward to your own chameleonic or immortality quotation submissions.

In *A Beginner's Guide to Immortality*, we'll journey through the lives of famous and obscure people, while wondering how their lives might have turned out differently if small events had changed during their development. Would Truman Capote ever have become a famous writer if his mother had not abandoned him, and he fled to New York City? Would John Cage have become the famous avant-garde composer if his mother hadn't told him to stay in Europe, when he wanted desperately to return to America? How would the world be different today if rocket-scientist Jack Parsons had never met the flamboyant redhead Marjorie Cameron or dropped a container of mercury fulminate in his private lab?

A Beginner's Guide to Immortality launches from person to person, touching on aspects of their lives that I find personally interesting. Will these chameleons be remembered a hundred years from now? For some, yes. Others, no. Either way, I hope to lengthen the impact of their lives for a few extra years before they, like each of us, are forgotten.

Through the centuries, many have striven to achieve "immortality" through science, myths, religion, or dreams of lifelike heavens—and also through a creative work that left some lasting mark. Let us recall the words of King Gilgamesh, in that epic Mesopotamian masterpiece written centuries before the Bible. Gilgamesh realizes that the only way for him to transcend death is to seek achievement, to do something beyond the traditional, to enter the mystical Cedar Forest and kill the monster Humbaba:

> We are not gods, we cannot ascend to heaven. No, we are mortal men. Only the gods live forever. *Our* days are few in number, and whatever we achieve is a puff of wind. Why be afraid then, since sooner or later death may come?. . . I will cut down the tree; I will kill Humbaba; I will make a lasting name for myself; I will stamp my fame on men's minds forever.[11]

🐾 🦚

I don't want to achieve immortality through my work; I want to achieve immortality through not dying.

—Woody Allen, *Without Feathers*

Life on Earth passes. It is not long. Being remembered is the only success.

—Advice to [Egyptian] King Merikare from his Father,
c. 1800 BC

As long as people talk about you, you're not really dead. . . . A legend doesn't die just because the man does.

—George Clayton Johnson, "A Game of Pool,"
Twilight Zone TV show, 1961

There is a theory that creativity arises when individuals are out of sync with their environment. To put it simply, people who fit in with their communities have insufficient motivation to risk their psyches in creating something truly new, while those who are out of sync are driven by the constant need to prove their worth. They have less to lose and more to gain.

—Gary Taubes, "Beyond the Soapsuds Universe"

A human being should be able to change a diaper, plan an invasion, butcher a hog, conn a ship, design a building, write a sonnet, balance accounts, build a wall, set a bone, comfort the dying, take orders, give orders, cooperate, act alone, solve equations, analyze a new problem, pitch manure, program a computer, cook a tasty meal, fight efficiently, die gallantly. Specialization is for insects.

—Robert Heinlein, *Time Enough for Love*

It has been said that something as small as the flutter of a butterfly's wing can ultimately cause a typhoon halfway around the world.

—*The Butterfly Effect*, movie

What are the [future office-place] skills that are going to be the most important for those kids? Is it going to be mastering new interfaces and keeping complex virtual relationships alive and multitasking and managing to think about new technologies in interesting ways? Or is it going to be algebra skills? I think you'd have to make the case that it's probably the former, not the latter.

—Steve Johnson, "Around the Corner," interview in *Time*, March 20, 2006

Pop culture comprises socially transmitted behavior patterns, art-works, and beliefs, the content of which is shaped by mass media such as movies and magazines. Pop culture evolves through a feedback loop between the media and its consumers. Sometimes, the feedback loop goes wild, and serious new trends are set, and fads are born. Religion is a singularity in such a loop.

—Cliff Pickover, *A Beginner's Guide to Immortality*

The chameleon's tongue shoots out of its mouth at more than 26 body lengths per second (13.4 miles an hour). It can snag prey located more than one and a half body lengths away. . . . The tongue accelerates from 0 to 20 feet per second in about 20 milliseconds—a rate so fast it defies the general principles of power production in muscles.

—Bijal P. Trivedi, "'Catapults' Give Chameleon Tongues Superspeed," http://news.nationalgeographic.com

A Beginner's Guide
to Immortality

ONE
TRUMAN CAPOTE AND *THE BRAIN FROM PLANET AROUS*
※ ♪

In which we encounter *The Brain from Planet Arous*, "The Visible Man," sex-starved alien brains, Truman Capote, the Fissure of Rolando, the stigmata of genius, *In Cold Blood*, Mia Farrow, *Music for Chameleons*, Candice Bergen, Norman Mailer, *Answered Prayers*, the grunting of a renegade hog, eccentric geniuses, the nature of creativity and intelligence, Blackwing-602 addicts, work habits of successful writers, homosexuality, mental disease, short people, guitarist Theodore Roosevelt Taylor, famous polydactyls, the "Black and White Ball," and the six-fingered Vladislav Khodasevich.

Failure is the condiment that gives success its flavor.

—Truman Capote[1]

Alien Brains

Last night, I dreamed I was eating the brain of Truman Capote. It tasted like black licorice, but it smelled of blood. Today, as I gaze out my loft window at crows in the grass, I can only wonder about the cause of my dream.

Capote, the audacious author of *In Cold Blood*, was one of America's most famous and praised writers. Alas, he died in 1984 after a long period of alcoholism that caused his brain to shrink. He now resides six feet under at Brothers Westwood Memorial Park in Westwood, California—next to other stars like Marilyn Monroe and Natalie Wood.

Cemeteries, brains, crows, and dreams—today, I feel like I'm in a Halloween novel. Outside, a crow squawks, and I turn my head in its direction. The black creature stares at me, its body motionless, its dark smile relentless.

I frequently have strange dreams, and in the latest dream, Capote's brain pulsated in my hand like a beating heart. I think I know what triggered my nightmare of necrophagy.

Earlier that evening, I had watched *The Brain from Planet Arous*, a movie from 1957, the year in which I was born. I had seen the movie in childhood on TV, and for years, as an adult, I wanted to see if the movie's floating brain and glowing eyes would still give me the shivers. The movie had a lasting impact on my life, triggering an obsession with human anatomy and biology. At age ten, the movie inspired me to purchase and assemble a plastic model called "The Visible Man," so I could impress people with the plastic brain, while pretending it was the alien brain from the movie. You can still purchase The Visible Man today with its endless, exciting features, which are advertised on the World Wide Web:

- Breastplate comes off
- Vital organs can be removed and replaced
- Translucent skin shows major veins and arteries
- Complete skeleton, eyes, and brain are included

The instruction manual suggested that vital systems might be painted to create a more realistic presentation. I always painted the organs with psychedelic colors—although I could never know the color of the Arous brain because the movie was not in color.

The Brain from Planet Arous is exciting, even with its weak special effects. But you're not exactly expecting a Steven Spielberg production from this low-tech 1957 movie, are you? Here are some movie highlights. An alien ship crash-lands in the California desert. The human protagonists spend time in a nuclear research lab with no obvious equipment. Later, they try to protect themselves against a sex-starved, alien brain who lusts for the beautiful leading lady played by Joyce Meadows.

Before we return to Truman Capote, let me tell you a bit more about *The Brain from Planet Arous*, which is available as a DVD today. Buy it—it will temporarily transport you to the 1950s, a time of Cold War paranoia during which the movie industry produced films about giant ants spawned by atomic tests, body snatchers, and invaders from outer space. Movies that today would be considered of low quality were capable of haunting audiences and giving them a sense that they swam in a sea of mystery. In a scene reminiscent of the violence of Capote's *In Cold Blood*, the alien brain meets its end when our human hero uses an ax on the brain's weak spot called the "Fissure of Rolando."

But I digress, as usual. My main interest for this chapter is Truman Capote, whose best-selling book features mayhem and carnage on a more graphic level than depicted in *The Brain from Planet Arous*. The movie inspired my lifelong obsessions with biology and fancy anatomical worlds like "fornix," "semilobar holoprosencephaly," "Fissure of Rolando," "medulla oblongata," "epoophoron," "galea aponeurotica," and "manubrium." Capote's book inspired me to be a writer.

Writing seemed like an impressive career path when I first heard of Truman Capote in 1965, when his famous *In Cold Blood* was published. Most likely, I saw my mother reading the book and grew curious about its subject matter (my dad never read books). *In Cold Blood* was Capote's carefully researched and frightening chronicle of a Kansas farm family that was slaughtered in 1959—followed by the eventual capture, trial, and execution of the two killers. At the time, I knew almost nothing about the book's content, but I did know that Capote took about six years to write it, and that it became a best seller, making Capote rich and famous. Maybe I could become famous too.

When Capote interviewed friends of the slain family, he took no notes. He simply stored what they said in his capacious mind and then dumped the contents of his brain into notebooks at the end of the day. Capote's chameleon-like ability to change colors allowed him to gain the confidence of his interviewees who ranged from Kansas housewives to F.B.I. agents. He pioneered a writing genre he called "the nonfiction novel," in which a true story is told with the pacing and dialogue of a novel. This technique allowed him to produce a journalistic, factual novel with the immediacy of film and the depth, artistry, and freedom of prose normally associated with fictional works.

The Life of Truman Capote

Truman Capote (1924–1984) said that he knew what his career would be from an early age: "Many people spend half their lives not knowing, but I was a very special person, and I had to have a very special life. I was not meant to work in an office, though I would have been successful at whatever I did. But I always knew that I wanted to be a writer and that I wanted to be rich and famous."[2]

Capote abused alcohol and drugs during the 1970s, and his health deteriorated. In his 1980 collection of nonfiction pieces, *Music for Chameleons*, Capote said that God had helped him in life. "But I'm not a saint yet," he wrote. "I'm an alcoholic. I'm a drug addict. I'm homosexual. I'm a genius. Of course, I could be all four of these dubious things and still be a saint."[3]

Capote, like me, had a particular penchant for making lists—some of my own favorite lists are presented in Chapter 8. For example, in *Music for Chameleons*, Capote enumerates the activities of which he was capable, including such accomplishments as: reading upside down, hitting a tossed can with a .38 revolver, driving a Maserati at 170 mph, and typing 60 words a minute. However, he listed some funny deficiencies as well. He couldn't recite the alphabet, couldn't subtract, and couldn't create a prepared speech—it always had to be spontaneous.

Capote frequently found himself isolated from his peers, even when growing up. "I was so different from everyone," he said, "so much more intelligent and sensitive and perceptive. I was having fifty perceptions a minute to everyone else's five. I always felt that nobody was going to understand me, going to understand what I felt about things. I guess that's why I started writing. At least on paper I could put down what I thought."[4]

Despite Capote's brilliance, many teachers considered him retarded. Their negativity, however, did not dampen his passion for writing. Of his early days, Capote said, "I began writing really sort of seriously when I was about eleven. . . . I used to go home from school every day and I would write for about three hours. I was obsessed by it."[5] His childhood psychiatrist acknowledged Capote's genius after observing him for a short time. However, Capote never completed high school and had no further formal education.

In Cold Blood starts in the poetic fashion that typifies Capote's works, "Until one morning in mid-November of 1959, few Americans—in fact, few Kansans—had ever heard of Holcomb. Like the waters of the river, like the motorists on the highway, and like the yellow trains streaking down the Santa Fe tracks, drama, in the shape of exceptional happenings, had never stopped there."[6] And then Capote gradually explores the mystery behind the brutal slaying of the Clutter family by two would-be robbers who held a shotgun a few inches from their faces. The murders had no apparent motive, and almost no clues existed. The frigid emotions of the killers were partly responsible for the book's appeal. "I really admired Mr. Clutter," said killer Perry Smith, "right up to the moment I slit his throat."

In order to write his book, Capote spent many hours interviewing Smith—learning about Smith's life, thoughts, and feelings. Several film reviewers of the 2006 movie *Capote* noted that "Capote struggled to find a balance between befriending Smith for the sake of the novel that will make him immortal, and truly convincing himself that he wants to save the man he has come to love."[7]

Capote flourished in his book's success and threw one of the 1960s most lavish parties—the infamous, exclusive, and legendary "Black and White Ball" at Manhattan's Plaza Hotel. Some called it the "party of the century." Capote required attendees to be chameleons and change their colors entirely—black and white evening clothes and exotic masks were mandatory. All ladies had to carry fans, and some guests commissioned exotic masks featuring interlocking swans or rabbit shapes made of mink. The dress code was so strict that the Secret Service agents accompanying first daughter Lynda Johnson had to dress appropriately. The guest list included America's power brokers in politics and the arts: "Those lucky enough to receive an invitation were, in Capote's and the world's eyes, 'the chosen'; those who were not were thrown into a panic of shameless striving. Some friends begged Capote for an invitation; others tried to bribe him with great sums of money. . . . Another was a woman who threatened to commit suicide if she wasn't invited to the party."[8]

Capote had come far from his humble beginnings in Monroeville, Alabama (population 1,600), where his mother sent him away to live with relatives because she was unsuited to raise a child. He always felt like an outcast. Who could have guessed he would someday hire the most elegant ballroom in America and have over five hundred of the most famous people in the world at his beck and call? Throughout the summer of 1966, Capote worked on his "guest list to end all guest lists," which included Frank Sinatra, Mia Farrow, Walter Cronkite, Walter Matthau, William F. Buckley, Jr., assorted Rockefellers, Candice Bergen, Lauren Bacall, Henry Fonda, Harper Lee, Norman Mailer, James Michener, Arthur Miller, Gregory Peck, Andy Warhol, and John Steinbeck.

All of Capote's fame came from essentially one book. After *In Cold Blood*, he received a large advance for a book to be titled *Answered Prayers*, a phrase that comes from Saint Theresa's assertion that answered prayers cause more tears than those that remain unanswered. He never finished this magnum opus due to his drinking and drug use. In 1982, physicians told him that his brain was literally shrinking due to his alcoholism and that he had only a few months to live. Capote seemed take the news calmly, saying, "Sometimes oblivion is a nice place to be."

Capote's childhood had been difficult. He was the son of a salesman, some say con-man, and a sixteen-year-old, discontented beauty queen, Lillie Mae Faulk. She desperately wanted to abort the embryonic Truman before he was born, but her husband said no. Soon after his birth, Lillie Mae and her husband parted ways.

A few years later, Lillie Mae sent her son away to live with "an eccentric family of maiden aunts and a bachelor uncle"[9] while she lived in a hotel with her new second husband. At the time, Capote said he felt like "a spiritual orphan." As we discussed, he was alienated from everyone he encountered due to his superior intelligence and perceptions.

Later, Lillie Mae regained custody of him, became alarmed with his effeminate mannerisms, took him to psychiatrists, sent him to military schools, and told her friends that she wished that she had given birth to a "dumb, football-playing" son instead of Truman.

We might speculate that Capote's lack of parental attention contributed to his desire for adoration and celebrity status in later years. Although he never went to college, he found work at the *New Yorker*, and attracted attention with his odd clothes that included a black opera cape.

In the last few years of his life, Capote slipped into a maelstrom of alcoholism, drugs, and destructive love affairs with younger men—and he seemed to find it impossible to complete *Answered Prayers*, which he had hoped would be his defining masterpiece. Finally, at age fifty-nine, he died of liver disease complicated by phlebitis and multiple drug intoxication. Some say the immediate cause of death was a drug overdose.[10]

The Stigmata of Genius

Few doubt Capote was an eccentric genius. Robert Linscott, his editor at Random House, said that "Truman has all the stigmata of genius."[11] Writer Norman Mailer summarized the view of many writers when he said that Capote wrote "the best sentences, word for word, rhythm upon rhythm" of any other writer.[12] Capote himself said, "To me, the greatest pleasure of writing is not what it's about, but the inner music that words make." Capote could also set the *mood* better than any writer of his time, as evidenced by this excerpt from "A Christmas

Memory," his story of an Alabama boy's connections to family and a vanishing American South:

> Morning. Frozen rime lusters the grass; the sun, round as an orange and orange as hot-weather moons, balances on the horizon, burnishes the silvered winter woods. A wild turkey calls. A renegade hog grunts in the undergrowth. Soon, by the edge of knee-deep, rapid-running water, we have to abandon the buggy. . . . "We're almost there; can you smell it, Buddy?" she says, as though we were approaching an ocean.[13]

Some of Capote's writing habits seem to be a little peculiar, but then again, many writers have had a touch of eccentricity. For example, Capote, like Mark Twain and Robert Louis Stevenson preferred to write while lying down—Capote described himself as a "completely horizontal writer."[14] Capote also did his best writing work in motel rooms. Similarly, other authors found it most productive to write while away from home—whenever George M. Cohan had a writing deadline, he bought a train ticket and spent the entire trip writing. According to *Writer's Digest*, he could "dash off 140 pages between New York and Chicago."[15] Like Thomas Wolfe, Capote preferred to write his books on yellow paper. Capote's favorite writing tool was the Blackwing No. 602, an extremely black lead pencil manufactured by Faber-Castell.

Many readers may wonder about Capote's love affair with the Blackwing-602; however, artists, designers, and writers continue to lust after the Blackwing. The world-famous pencil collector Doug Martin (www.pencilpages.com) says that he is drowning in queries from former Blackwing-602 addicts who are looking for the pencils.[16] The habit-forming Blackwing is discussed in online forums, and has been the subject of newspaper articles. Martin notes that buyers and sellers use eBay and his own Web site to transact Blackwing deals, with prices exceeding $20.00 per pencil.[17]

Blackwing-602 devotees spend $250.00 for a box with a dozen of

these upscale pencils, because they allow a writer to smoothly produce an ultrablack line with "half the pressure, twice the speed."
Sadly, the Blackwing-602 was discontinued in 1998 and has not been
made since. Martin, and other pencil lovers, have contacted the manufacturer, requesting that Blackwing production be resumed. He
writes, "There has been talk of circulating a petition to try to persuade
the company to fix the [Blackwing-602 production] machine. Well, I
have bad news. . . that machine no longer exists. It was scrapped to
make room for other equipment. There will never be another Blackwing 602 in its popular form."[18]

According to David Weeks and Kate Ward, authors of *Eccentrics*,
unusual people like Capote are essential in a society because such
novel points of view offer the variety a culture needs to adapt successfully to changing conditions. "Eccentrics may fail in any single
endeavor," they write, "but society wins by their examples and by what
can be salvaged from the ideas, problems and questions which they
radiate."

Perhaps the peculiarities, or even physical or mental defects exhibited by many geniuses, have caused these individuals to overcompen-
sate through constant creative activity. For example, in studying
eccentric geniuses, we find that many have had a sense of physical
vulnerability and the existence of a psychological "unease," that at
any moment they could be without a means of income, or incapaci-
tated or sick, or could die. Perhaps this unease keeps individuals on
edge and serves as a source of creative tension. Their works often bear
a personal signature, and a striving for dominance or power.

In the scientific arena, many cutting-edge geniuses had to persevere
despite resistance. For example, Scottish biologist Alexander Fleming's
revolutionary discoveries on antibiotics were met with apathy from his
colleagues. Many surgeons initially resisted English surgeon Joseph
Lister's advocacy of antisepsis. American inventor Chester Carlson,
inventor of the Xerox machine, was rejected by more than twenty companies before he finally sold the concept.

As I point out in my books *Strange Brains and Genius* and *Sex, Drugs, Einstein, and Elves*, through history, geniuses like Capote not only differed from the norm in intelligence, but also in some other mental or physical characteristic. For example, Alexander, Aristotle, Archimedes, Attila, John Hunter, William Blake, and St. Francis Xavier were all short (most less than five foot two). Capote was a little over five feet tall and had a high, lisping voice and effeminate manner.

According to Arnold M. Ludwig, author of *The Price of Greatness*, many great people have had a deformity of one kind of another: Susan B. Anthony (crossed eyes), Vladislav Khodasevich (six fingers on each hand), Jane Addams (spinal disfigurement), Ring Lardner (deformed foot), Allen Dulles (club foot), and Claude Debussy (large bony protuberances on his forehead). Aristotle, Aesop, Demosthenes, Virgil, Darwin, and Henry Cavendish all had a stammer.[19] The famous composer Gustav Mahler had an obsession with his mother that may have manifested itself in a limp that he unconsciously adopted to simulate his mother's lameness.[20] Many geniuses had asthma, incapacitating allergies, obsessive compulsive disorder, or bipolar disorder. Could it be that chronic physical ailments give some individuals the desire to compensate for their shortcomings, or to leave a mark on the world and achieve immortality through creative excellence?

Recent studies suggest that bipolar disorder and creativity share a common root. For example, Kiki Chang and Terence Ketter of Stanford University studied the children of parents afflicted with bipolar disorder (also known as manic depression). Some of the children actually suffered from the disorder, while others were at high risk of developing it. On a creativity test measuring responses to line drawings, *both* groups of children outscored children who were not born to bipolar parents, suggesting that full-blown bipolar episodes are unnecessary for creativity, but rather that creativity and bipolar disorder may stem from a common biochemical or psychological cause.

Polydactyl People and the Quest for Greatness

As a curious aside, famous polydactyls—people born with more than the usual number of fingers or toes—include Charles VIII of France, Winston Churchill, and Pacal the Great (King of Palenque), and as mentioned in the previous section, Russian poet Vladislav Khodasevich. Biblical giants were said to exhibit sexdactyly, a genetic condition in which a person has six fingers or toes:

> And there was yet a battle in Gath, where was a man of great stature, that had on every hand six fingers, and on every foot six toes, four and twenty in number; and he also was born to the giant. And when he defied Israel, Jonathan the son of Shimeah the brother of David slew him. (2 Samuel 21)

My favorite creative polydactylic is Theodore Roosevelt Taylor (1915–1975), the jazz guitarist born with six fingers on each hand. Like Capote, he was abandoned by his parents in his childhood. One day, Taylor's stepfather stood in the doorway with a shotgun, and told Taylor to get the hell out. In his early twenties, Taylor played guitar and piano along the Mississippi Delta. In 1942, the Klan chased him out of Mississippi after he had an affair with a white woman. Taylor spent the first day hiding in drainage ditches and then headed for Chicago where he became one of Chicago's most famous bluesmen. I have never been able to ascertain whether his six fingers affected his guitar playing ability, but I do know that while drunk, he attempted to cut off the extra fingers. At the height of his recording career, he still often preferred to use his $50 Japanese electric guitar and Sears Roebuck amplifiers with cracked speakers to get the distortion he so loved. You can listen to snippets from his music at www.amazon.com.

Note that a full complement of fingers is not necessary to soar with the guitar. For example, guitarist Django Reinhardt (1910–1953) astounded jazz lovers with his incredible command of the guitar, even with his deformed hand. After a fire that left him with the use of only two fingers on his left hand, he taught himself a fingering system that allowed him to rapidly play the most complex songs that most "normal" guitarists could never hope to attempt. As some reviewers have exclaimed, "Django could play more with two fingers than anyone could play with twenty."

Surprisingly, finger lengths are markers of your entire personality and career choices! Recent studies show that the shorter the index finger is compared to the ring finger, the more aggressive a man is likely to be. The longer a man's fingers (especially the ring finger) are relative to his height, the more likely he is to suffer from depression. Women with shorter ring fingers than forefingers tend to have higher fertility, and men with relatively long ring fingers are more likely to have high fertility.

A study of over one hundred male and female academicians at the University of Bath found that ring finger ratios could be used to predict tendencies for a man to teach in hard sciences (including mathematics and physics) versus social sciences (psychology and education). For example, men with index fingers as long as their ring fingers tended to be in the hard sciences. Men with ring fingers longer than index fingers tended toward the social sciences. Studies of male and female students taking computer classes showed that the smaller the difference is between index and ring finger lengths, the higher the overall test score. All of these results appear to be statistically significant! Apparently, relative ring finger length is also correlated with estrogen and testosterone levels.[21]

Genius, Sexuality, and Offspring

Returning to the subject of genius, very few brilliant people have given rise to exceptional progeny—either because great geniuses were celibate, or if married didn't want children, or because the children of geniuses were rarely geniuses. Francis Bacon once wrote, "The care of posterity is most in them that have no posterity." The great English poets or essayists Shakespeare, Johnson, Milton, Pope. Dryden, Gold-smith, Shelley, Keats, and Addison did not produce exceptional children. Many male geniuses were distinctly asexual or homosexual, and never married. Michelangelo, for example, said, "I have more than enough of a wife in my art." Electrical genius Nikola Tesla's said, "I do not think you can name many great inventions that have been made by married men." Other notable celibates or homosexuals include: Leonardo, Newton, Kant, Galileo, Descartes, Spinoza, Florence Nightingale, Copernicus, Handel, Cavour, Flaubert, and Chateubriand. Philosopher Kierkegaard was celibate and considered sexual relationships an abomination. Saint Paul preferred sexual abstinence. Tennyson was asexual, never having kissed a woman until, and at age forty-one, he married a thirty-seven-year-old invalid with a spinal problem. When Flaubert's seizures started, he gave up sex for a year and then developed homosexual interests.[22]

Clearly, many great, influential people fall outside the "norm" when it comes to sexuality, at least according to the Wikipedia encyclopedia.[23] Example homosexuals or bisexuals include Jane Addams (American social reformer), Alexander the Great, (Macedonian King; bisexual), W. H. Auden (British poet), Francis Bacon (British philosopher and scientist), James Baldwin (American author), Leonard Bernstein (US composer and conductor; bisexual), John Cage (American composer and partner of Merce Cunningham), Truman Capote (American author), Marcel Duchamp (artist), Allen Ginsberg (Beat poet), Hadrian (Roman military commander and emperor), Dag Hammarskjöld (former Secretary-General of the United Nations), John Maynard Keynes (British economist), Liberace (pianist), Socrates (Greek philosopher), Solon (Greek statesman), Gertrude Stein (American author, partner of Alice B. Toklas), Alan Turing (British mathematician, computer scientist), Andy Warhol (American artist and pop art icon), Mae West (American actress), Edmund White (American novelist), Walt Whitman (American poet), Oscar Wilde (Irish playwright), Thornton Wilder (playwright), Tennessee Williams

(American playwright), Ludwig Wittgenstein (philosopher), and Virginia Woolf (British author). Of course, lists such as these do not tell us if a statistically significant percentage of homosexual, asexual, or bisexual people are more creative or intelligent than heterosexuals.

Dr. James Weinrich of the University of California, San Diego has investigated the relationship between various forms of nonreproduction (especially homosexuality) and intelligence as measured by IQ and other tests. His published findings suggest that homosexuals score higher on intelligence tests than heterosexual control subjects. (The only exceptions to this trend are concentrated in prisoner populations.) He writes, "Moreover, the more representative the sample studied, and the less subject to challenge the methodology used, the clearer and more statistically significant was the superiority in intelligence of the more homosexual over the more heterosexual group."[24] Arnold Ludwig, Professor of Psychiatry at the University of Kentucky Medical Center, finds marked differences in sexual orientation among various professions—the creative arts professions attract higher proportions of homosexuals and bisexuals; the sciences attract (or promote) a higher proportions of hyposexual (sexually abstemious) persons; and the "enterprising" professions, such as politics, business, and the military attract the highest proportions of heterosexuals.[25]

Richard Florida, author of *The Rise of the Creative Class,* suggests that even if homosexuals are not more creative than heterosexuals, their occurrence in a community is conducive to creativity. Their presence in large numbers is an "indicator" of an underlying culture that is tolerant and diverse. According to Florida, American cities that include a large gay community are also "high-tech hotspots" and have strong economic growth.

Genius, Precociousness, and Suicide
Like Capote, most geniuses display their special minds from early ages. Mozart was playing the piano at three. Francis Galton and

Jeremy Bentham knew Latin and Greek at the age four. The Italian poet Torquato Tasso spoke at six months and studied grammar at three years. Scientist Lord Kelvin entered Glasgow University at ten. Also, just like Capote, many great geniuses performed poorly at school or dropped out. Such "dunces" included Isaac Newton—who later in life also suffered from nervous breakdowns—and physicist James Maxwell. Albert Einstein's teacher said that his "presence in the classroom is disruptive and affects the other pupils." Many influential geniuses dropped out of school. Electrical genius Oliver Heaviside went to school until age sixteen and had no formal education afterward. Thomas Edison, the Wright brothers, George Stephenson, Charles Dickens, Mark Twain, and Maxim Gorky did not complete high school. Presidents Washington, Lincoln, and Truman never earned a college degree.[26]

Ludwig has found that about one-third of the eminent poets, musical performers, and fiction writers suffered from serious psychological problems as teenagers, a rate that exploded to three-quarters when these people became adults. Author A. J. Jacobs comments on the endless parade of self-destructive writers: "A French writer hanged himself from a lamppost. A Peruvian did it in a deserted classroom. A Japanese poet finished himself off with his mistress at a mountain retreat. They throw themselves down stairwells, they leap from bridges. One Hungarian writer weighted his clothes down with stones and jumped in a lake, foreshadowing Virginia Woolf."[27]

Virginia Woolf was one of the greatest innovators in the English language, particularly famous for her stream-of-consciousness writings and for exposing the deep emotional motives of her characters. In 1941, Woolf filled her pockets with stones, and drowned herself in a river. The suicide note for her husband read:

I feel certain that I am going mad again: I feel we can't go through another of those terrible times. And I shan't recover this time. I begin to hear voices, and can't concentrate. So I am doing what seems the best thing to do. You have given me the greatest possible happiness. . . . I can't fight it any longer. . . .[28]

Writers John Berryman, Sylvia Plath, Anne Sexton, Ernest Hemingway, and Hunter S. Thompson all committed suicide. Alcoholic Berryman jumped to his death off the Washington Avenue Bridge in Minneapolis. (Berryman's father had also committed suicide.) Sylvia Plath took a bottle of sleeping pills and stuck her head in a gas oven. Anne Sexton committed suicide by carbon monoxide poisoning. Hemingway died from his shotgun blast to the head. Hemingway's father had also committed suicide—using an antique era pistol. Hemingway's granddaughter, Margaux Hemingway, killed herself by overdosing with Klonopin, a drug described for panic disorder. Hemingway's brother and sister also killed themselves.

In 2005, Tristan Egolf, age thirty-three, killed himself with a shotgun. Egolf, a novelist whose prose has been compared to the work of Thomas Pynchon and William Faulker, is famous for *Lord of the Barnyard*, a 410-page novel without any dialogue. The book was rejected by more than seventy US publishers before it was published.

Japanese author (and homosexual) Yukio Mishima liked to pose as a dead person in photographs depicting drowned sailors, saints shot with arrows, or a samurai committing ritual suicide. He ended up disemboweling himself with a knife.

This astonishing list of suicidal writers does not include the long list of authors who *tried* to kill themselves but failed: Joseph Conrad, Maxim Gorky, Guy de Mupassant, Eugene O'Neill, and many more. Science-fiction novelist Philip K. Dick tried to kill himself by taking seven hundred milligrams of potassium bromide.[29] Jack Kerouac died from an internal hemorrhage at the age of forty-seven, the result of a life of heavy drinking and drug use.

In fact, many of the writers that teachers told me to read while in college and high school were drunks: Truman Capote, Jack Kerouac, F. Scott Fitzgerald, Sinclair Lewis, Raymond Carver, Adela Rogers St.

John, Jean Stafford, Edgar Allan Poe, Stephen Crane, Theodore Roethke, Herman Melville, Delmore Schwartz, William Faulkner, Ernest Hemmingway, Allen Tate, Caroline Gordon, Ring Lardner, Dorothy Parker, Robert Lowell, Eugene O'Neill, John O'Hara, O. Henry, Conrad Aiken, John Berryman, Edmund Wilson, and Jack London. London committed suicide at age forty.

According to psychiatrist Donald W. Goodwin, author of the *Alcohol and the Writer*, statistics show that more writers die of cirrhosis of the liver, a disease closely associated with alcoholism, than people in other occupations (not counting bartenders).[30] Seventy percent of the Americans who have won the Nobel prize for literature were either alcoholic or drank to excess. (We'll talk about drug-using writers in Chapter 6.) I have not been able to determine *why* an apparent link between creativity and alcoholism exists. Is it possible that writers spend many hours alone while working on a book, and this isolation favors drinking? Because writers are "self-employed," perhaps they can afford to be drunk, whereas people in other professions would be fired. Another theory is that alcohol is a "socially acceptable" self-medication for bipolar disorder—and bipolar disorder is known to be prevalent in the highly creative. Could alcohol increase literary creativity levels in some people? Many unanswered questions exist.

Recently, scientists have discovered that they can predict with high likelihood which poet is likely to attempt suicide by analyzing words in the poet's writings. According to a study published in *Psychosomatic Medicine*, and supported by a grant from the National Institute of Mental Health, poets who ultimately committed suicide: 1) used more words associated with death than did the non-suicide group; 2) used more first-person singular self-references (such as "I," "me," and "my"); 3) used fewer first-person plural words than did the nonsuicidal poets; and 4) used fewer communication words (such as "talk," "share," and "listen").[31] We do know that suicide rates are much higher among poets than among other literary writers and the general population, although most poets do not commit suicide.[32]

American poet Hart Crane (1899–1932, famous alcoholic homosexual) killed himself by leaping into the ocean. Russian poet Sergei Yesenin (1895–1925, alcoholic) slashed his wrists and wrote his epitaph with his own blood: *"Do svidan'ia, drug moi, do svidan'ia'* (In this life it is not new to die, / but neither it is new to be alive)." Incidentally, Yesenin's life has echoes of Capote's—Yesenin, the child prodigy, was abandoned by his parents and he began writing impressive poetry at age nine.

Australian poet Adam L. Gordon (1833–1870) killed himself and left behind this gem: "Life is mostly froth and bubble / Two things stand like stone / Kindness in another's trouble / Courage in your own." American poet Randall Jarrell (1914–1965) attempted suicide by slashing his wrist. Later, while at a hospital receiving wrist therapy and while walking at dusk on a nearby highway, a car struck and killed him. He had written, "A good poet is someone who manages, in a lifetime of standing out in thunderstorms, to be struck by lightning five or six times."

American poet Sara Teasdale (1884-1933) overdosed on sleeping pills. Russian poet Vladimir Mayakovsky (1893-1930) shot himself, even after he had condemned the suicide of the poet Sergei Yesenin. In his suicide note, Mayakovsky wrote, "Mother, sisters, friends, forgive me—this is not the way (I do not recommend it to others), but there is no other way out for me. Lily—love me." His funeral procession and ceremony drew thirty thousand people. In 1930, his birthplace was renamed to Mayakovsky in his honor. (Rumors arose that Mayakovsky did not commit suicide but was, in fact, murdered at the behest of Stalin; however, this hypothesis remains unsubstantiated.)

Pulitzer-prize winning poet Robert Lowell (1917–1977) is a final spectacular example of mental disorder in poets. Lowell was a heavy drinker, hospitalized dozens of times for manic depression, and referred to one of his manic episodes as a "magical orange grove in a nightmare." Of Lowell's life, novelist Walter Kirn writes: "Lowell's poems proved that if writing is a form of therapy, it's a uniquely unsuccessful one, at least in medical terms, and that insights into the larger human predicament don't guarantee their author a good night's sleep, a stable marriage or a dignified passing. Winning Pulitzer Prizes and the like is no balm either. Nothing (even lithium, it seemed) could halt Lowell's slide into miserable ill health and psychological chaos."[33]

The time Lowell spent in mental hospitals was occasionally productive. From his padded cell in Baldpate Hospital in Georgetown, Massachusetts, he wrote in a letter, "I'm in grand shape. . . . The world is full of wonders." He composed his famous love poem "Walking in the Blue" while a patient at McLean's Hospital.

In 1974, Lowell wrote to poet Elizabeth Bishop: "I see us still when we first met. . . . I was brown haired and 30. . . . I was largely invisible to myself, and nothing I knew how to look at. But the fact is we were swimming in our young age, with the water coming down on us, and we were gulping. I can't go on."[34]

Recent research demonstrates that having a quirky approach to life may be one of the keys to becoming a great artist, composer, or inventor. For example, individuals with schizotypal personalities are more creative than either normal of fully schizophrenic individuals. Psychologists have a range of tests to diagnose schizotypal personality disorder, and define it as being characterized by peculiarities of thinking, odd beliefs, and eccentricities of appearance, behavior, interpersonal style, and thought. Brain scans show that schizotypes rely more heavily on the right sides of their brains than the general population to access their creativity.[35]

In order to demonstrate their theories of schizotypes, Vanderbilt University psychologists Brad Folley and Sohee Park had schizotypes, schizophrenics, and normal individuals invent new functions for various objects. The schizotypes suggested creative new uses for the objects much more often and more quickly than either of the other categories of people. Near-infrared optical spectroscopy (brain scans) performed while the subjects invented showed that the schizotypes' right hemispheric activation was dramatically greater than exhibited by the other subjects.[36] Swiss neuroscientist Peter Brugger has also shown that the right hemisphere plays a crucial role when people are making novel associations.

Highly creative people do appear to have more sex than average. According to a 2005 survey by British researchers Daniel Nettle and Helen Keenoo, professional artists and poets have around twice as many sexual partners as other people. Volunteers were also assessed for character traits associated with schizophrenia. Several professional artists and poets scored as highly on these measures as actual schizo-phrenics! Researchers conjecture that some of these traits, such as mag-ical thinking, are linked with increased sexual activity and evolved because they contribute to the survival of the human species. Research is being conducted to ascertain if the more creative a person is, the more sexual partners the person is likely to have.

Obviously, not all creative people had extreme psychological or physical problems, or exhibited homosexual or celibate behavior—and not all people with such characteristics are very creative. We can offer long lists of fairly "normal" artists, writers, and musicians. Neverthe-less, I've found that an unusually high number of creative people, par-ticularly in the arts, have suffered from mental disorders of one kind or another. Many were far from the norm in many areas of their lives. About Capote, Christopher Lehmann-Haupt writing in the *New York Times* sums up nicely:

> The pieces in *Music for Chameleons* have freed Capote to write about himself—even to confess, without a trace of self-pity or bravado, the agony he felt as a child over his secret desire 'to be a girl.' Yet these pieces can hardly be called an egotistical celebra-tion of his personality. He does what he does with art. That art is a sort of music. We gather to listen and to blend ourselves into the composer's background. Just like the chameleons.[37]

⚜

Life is a moderately good play with a badly written third act.
                                                                    —Truman Capote

Be like the chameleon. Change colors. Be flamboyant. Look in
all directions. Dream.
                                            —Xyla Bone, *I Am My Husband's Hippocampus*

And, indeed, it is a kind of ocean. Scented acres of holiday trees,
prickly-leafed holly. Red berries shiny as Chinese bells: black
crows swoop upon them screaming. Having stuffed our burlap
sacks with enough greenery and crimson to garland a dozen
windows, we set about choosing a tree.
                                            —Truman Capote, "A Christmas Memory"

All depressed people share the feeling that they are unconnected
to others. Creativity is a way to connect while staying safely self-
encased. For the depressed, creativity is therapy.
      —Leslie J. Miller, "There's Nothing Deep About Depression,"
                                                New York Times Magazine, 2005

Capote demonstrated that reality, if heard out patiently, could
orchestrate its own full range. He did not intend to be merely the
novelist-as-journalist writing diversionary occasional pieces. He
had already done all that in "Local Color." In the completer role
of novelist-as-journalist-as-artist, he was after a new kind of statement. He wanted the facts to declare a reality that transcended
reality.
          —Conrad Knickerbocker, "One Night on a Kansas Farm,"
                                            The New York Times, January 16, 1966

A single gene in the fruit fly is sufficient to determine *all* aspects of the flies' sexual orientation and behavior [homosexuality and lesbianism].

> —Dr. Barry Dickson in "For Fruit Flies, Gene Shift Tilts Sex Orientation," *The New York Times*[38]

The unusual pattern of genetic diseases seen among Jews of central or northern European origin, or Ashkenazim, is the result of natural section for enhanced intellectual ability [due to persecution in Medieval Europe]. Ashkenazi Jews make up 3 percent of the American population but won 27 percent of its Nobel prizes, and account for more than half of world chess champions.

> —Nicholas Wade, "Researchers Say Intelligence and Disease May be Linked in Ashkenazic Genes," *The New York Times*[39]

What is meant by "reality"? It would seem to be something very erratic, very undependable—now to be found in a dusty road, now in a scrap of newspaper in the street, now in a daffodil in the sun. It lights up a group in a room. . . . It overwhelms one walking home beneath the stars. . . . Sometimes, too, it seems to dwell in shapes too far away for us to discern what their nature is. But whatever it touches, it fixes and makes permanent. That is what remains over when the skin of the day has been cast into the hedge. . . .

> —Virginia Woolf, *A Room of One's Own*

TWO
JOHN CAGE AND THE ZEN OF MUSIC
❧ ❧

In which we encounter John Cage, the future of music, the end of movies, "Europera 5," the mystery of silence, "Eclipticalis With Winter Music," amygdala-stimulation movies, neoproterozoic Lake Vostok, deadly mushrooms, overspecialization, Johann Wolfgang von Goethe, premaxillary bones, Samuel Johnson, Haydn's *Farewell Symphony*, Francis Galton, the Biblical book of Job, and the silence of God.

If something is boring after two minutes, try it for four. If still boring, then eight. Then sixteen. Then thirty-two. Eventually one discovers that it is not boring at all.

—John Cage

Brain Symphony

The movie *The Brain from Planet Arous* did have its shortcomings. For example, the visual special effects were sometimes so simple that movie producers made little effort to hide the wires that lifted the floating brains. When our hero strikes the brain, it bobs up and down on a guide-wire like a piñata tied to a ceiling beam. Nevertheless, because the scenes have so much action, wrinkled brain material, and psychological tension, many viewers overlook the technical shortcomings.

The music for the movie, composed and conducted by Walter Greene (1910–1983), is haunting, and reminiscent of the excellent sound track for the original *The Outer Limits* TV series. In fact, *The Brain from Planet Arous* seems incomplete and lifeless without the music to heighten the suspense and to suggest the alien brain's moods.

In the future, when we interface directly to movielike entertainment through neural connections, perhaps music will be unnecessary. A movie director will be able to shape your mood simply by tickling the appropriate neural nets near the amygdala, a small almond-shaped structure deep inside the brain. For example, a portion of the amygdala known as the lateral nucleus appears to play a key role in fear. Instead of superimposing the sounds of a screeching violin over a scary scene, the movie director will stimulate the amygdala, which will automatically generate signals to trigger autonomic arousal, such as a faster heartbeat and the production of stress hormones. When you see the floating brain on the screen, you will feel "real" fear. In other movies, you'll feel a state of bliss when you seen an angel or sexual arousal when the lead actor comes on stage.

Even if brain interfaces gradually emerge, music will remain important in entertainment for many decades. In the future, when separate brains interconnect by wireless transducers, multiple musicians and viewers will share rhythmic and melodic motives so that members of a collective mind can develop one another's inputs to form large-scale symphonies. Eventually, animals or artificial entities will contribute material to the hypermusic, selecting musical fragments and layering them onto the symphony in the same way that an artist dabs on paint. While listening to music, the brain's pleasure centers will be stimulated at particular times, which may be determined by voting among members of the collective.

Twenty years after music becomes unnecessary in movies due to direct emotional stimulation, movies themselves will begin to disappear. When we enter virtual worlds in computerlike devices, we will have our own adventures instead of watching the adventures of others. We won't observe movie actors for excitement or to escape from our ordinary lives. Instead, we will participate directly in the journey and have the virtual experience of climbing Cheop's pyramid, scuba diving in neoproterozoic Lake Vostok, or exploring the Brazilian rain forest while being accompanied by Julius Caesar or a scantily clad Cleopatra.

A dial will allow us to determine the degree of realism, immersion, and sensory impact.

A hundred years after the post-movie era, we may lose adventure entertainment entirely and exist in a state of higher consciousness and bliss. We will no longer crave relentless action such as car chases, gun shootings, or shark attacks. How many movies do you think the Dalai Lama or any other holy person needs to watch? How many James Bond movies does a monk studying the Ambrosian Rite in the Church of Milan want to see? According to Canadian pianist and essayist Glen Gould,

> In the best of all possible worlds, art would be unnecessary. Its offer of restorative, placative therapy would go begging a patient. The professional specialization involved in its making would be presumption. . . . The audience would be the artist and their life would be art.[1]

In *The Brain from Planet Arous*, the advanced beings have no need of bodies. In our own future beyond the post-music age, whether we are uploaded to virtual worlds or are still present in this one, we will be able to induce a feeling of stillness, serenity, a sacred presence, as if we are on the threshold of transcending space and time. Person-to-person communication will be pervasive and instantaneous. Each one of us will serve as a nucleation site around which form the crystals of spiritual expression that lifts a totally interconnected society out of purely pragmatic endeavors and places us in direct contact with the numinous. The notion of "I" will periodically dissolve; the thinking mind will turn off, but we will still be conscious. There will be no suffering; we'll enter a timeless state in a seemingly endless "now."

Even today, researchers like Warren Meck and Catalin Buhusi of Duke University in Durham, North Carolina, are investigating how the brain monitors the passing of time via a region of the brain called the striatum. Could a tiny alteration of the striatum give us a simulacrum of immortality? In the future, all of us may be able to alter our experience of time by using designer drugs or by manipulating the brain's dopamine system, which is known to alter temporal perception. According to Robert Levine of California State University in Fresno, "Time is our most valuable possession. Until the biomedical people can make us live forever, the closest thing we have is to stretch the moment."

Chameleon Cage

Let us return from this wild speculation and focus on current times. In the early twenty-first century, we do have music, and music helps us transcend our everyday lives. As avant-garde musician John Cage once suggested, "Music is a way of preparing our minds and our personalities—to finally throw away music altogether and experience the world directly."[2]

John Cage (1912–1992) was a chameleon like Capote, with many interests, talents, and a sense of the flamboyant. Not only was Cage a musical composer, writer, and philosopher, he was a passionate mycologist and mushroom collector. He started a musical revolution when he tore down an entrenched musical superstructure that had evolved since the time of Beethoven. His works were minimalist, making us smile in the same way we smile at the bare-bones special effects in *The Brain from Planet Arous*. Cage didn't care if the musical "guide-wires" showed as he made his statements to the world. The music critic's opinion didn't count for much. The *act* of composing wasn't important. It was the final *result* that mattered. As we discussed, Truman Capote once said, "To me the greatest pleasure of writing is not what it's about, but the music the words make." I'm sure Cage would agree. The precise "meaning" of Cage's composition is sometimes tenuous, but he reveled in the strange aural results and the music of quiet chaos.

Some of Cage's works were conventional, but he is famous for his avant-garde compositions designed for every imaginable and unimaginable kind of instrument. He occasionally employed special pianos supplemented with paper, wood, and rubber bands placed between the strings to make the piano sound alien. His musical scores sometimes called for sounds made by toys, cucumbers being chopped, flower pots, cowbells, and electronic oscillators.

Cage's famous *Europera 5* was his last opera, and it employed a collage of opera singers, a 78-rpm Victrola record player, a prerecorded tape, and a television and radio broadcast. The acoustical result is rather "three-dimensional" and appears to fill a very large space. Listeners can only perceive the modern radio broadcasts during the silences of the opera. Imagine for a moment that you were in the audience when the piece premiered on April 12, 1991, at Slee Hall, State University of New York, Buffalo. On stage are two singers, each singing five arias of their own choosing. A pianist plays six different opera transcriptions. A *silent* television is also on stage.

Today, some of us may wonder why there is so much fuss about "Europera 5." Was it a great work or utter crap? Critics claim that the separation of the various operatic elements produces a "spaciousness and awareness of distances that is so characteristic of Cage's music." A recording of "Europera 5" is available for purchase on the Internet, and the reviews are good. Damon Krukowski of *Tower Classical Pulse!* suggests that this is like "no opera you have ever heard, and yet it is every opera you have ever heard. . . . It refocuses your attention on familiar sounds that then become completely new in the context of the piece."[3] Art Lange, writing for *Fanfare*, notes, "There's a dreamlike ambiance, a marvelous confusion of time and plot, as in the 'duets' of different material between the live and 'historic' singers."[4]

The two singers move along a grid divided into sixty-four squares. Cage's directions ensure that whenever more than one of the three musicians is performing, the others are in the background. This helps prevent the production from becoming pure chaos. Joan La Barbara, writing for *Schwann Opus*, notes:

By allowing singers to choose their favorite arias, sung *a cappella* or against unrelated coincidental accompaniment, Cage presented them unfettered, juxtaposed, out of sequence, out of time and space, as individual gems to be rethought. Like ghosts of operas past, fragments of classic 78-rpm recordings emanate from a vintage Victrola. [The taped superimposed operas] offer filmy glimpses of bygone performances set against living voices performing their showy war horses. . .[5]

Let's discuss more famous and cherished Cage compositions. After spending some time in an anechoic chamber at Harvard University and observing "ultimate silence," he composed his most famous and controversial piece titled "4'33''"—4 minutes and 33 seconds of silence, divided into three movements!

At the 1952 premiere, which featured pianist David Tudor at Woodstock, New York, some listeners were unaware that the piece was finished when it had concluded. During the performance, Tudor actually placed a handwritten score, with blank measures, on the piano and sat motionless. Cage and Tudor both felt that the score was essential to the performance, alerting the audience to the fact that something was happening.

During the piece, the audience whispered to one another, and some people left. When Tudor finished performing "4'33''," raising the keyboard lid and standing, some in the audience screamed in an uproar. A number of audience members were not pleased.

Earlier, when Cage had visited the Harvard anechoic chamber, he had expected to hear nothing. However, he said he perceived low and high sounds. He was told that the first somehow related to impulses of his nervous system. The second was the sound of his blood circulating. This experience taught him that silence never really existed—and during "4'33''", the audience heard phenomena like whispers, room noises, passing traffic, and the wind blowing in the nearby trees. That was the point of "4'33''"—to become aware of all the unpredictable sounds of silence that become the music of the piece.

"4'33''" has since been performed numerous times around the world. In 2002, John Cage's publisher filed plagiarism charges against

Mike Batt—one of Britain's best-known songwriter/composers—after Batt published an album titled *Classical Graffiti* and credited his own silent song "A Minute's Silence" to "Batt/Cage." Batt settled the lawsuit out of court for an undisclosed six-figure sum.

Previous musicians have experimented with music that fades to silence in interesting ways. For example, in Joseph Haydn's Symphony No. 45, written in 1772, the musicians one by one stop playing and walk from the stage. Also known as the *Farewell Symphony*, the only remaining musicians at the end are two muted violins. Together, the violinists played the symphony's last note, then blew out nearby candles, and left. The *Farewell Symphony* was Haydn's subtle way to remind his sponsor, Prince Nicholas of Esterházy, that the musicians were tired of playing and anxious to leave the Prince's summer palace to return to their wives in Vienna.

Another favorite example of avant-garde musical composition includes the work of American composer La Monte Young, a pupil of John Cage and fellow "minimalist" who produced Composition 1960 #5, in which the performer is instructed to turn butterflies loose in the performance area. The piece is finished when the butterflies have flown away.

Perhaps a century from now, others will be toying with the brain as Cage did with music. Instead of turning off the sound, we'll turn on and off brain regions. Musical scores will indicate temporal regulations of areas such as the corpus callosum, cingulate gyrus, hypothalamus, and cerebral peduncle. Imagine an attractive percussionist staring at you, with his or her drums and cymbals networked to your brain. The cymbals elicit a mild syncopation of your prefrontal cortex and thalamus, thus driving you to an exquisite exallotriote ecstasy that eclipses reason and ordinary emotion.

Eclipticalis with Winter Music

As you might expect, traditional musicians often dismissed Cage as a nut. During the 1964 New York Philharmonic performance of "Eclipticalis with Winter Music," nearly half of the audience left, and the string section of the orchestra booed Cage. To compose the piece, Cage superimposed musical staves over star charts in an atlas of stars. Brightness of the stars was translated into the size of the notes.

Although too creative for some, Cage was certainly a chameleonic genius—he had graduated from Los Angeles High School as class valedictorian. Later he taught at the University of Southern California in Los Angeles.

In the 1950s and '60s, Cage used the *I Ching* and tossed coins to determine the rhythms and pitches for his pieces. In 1952, he composed a dance piece by recording forty-two phonograph records on tape, chopping the tapes into short pieces, and using chance operations to determine how the tape fragments should be reassembled. How would you like to dance to that?

In 1969, he composed "Hpschd" (1969), a collaboration with Lejaren Hiller for seven harpsichords, fifty-one tapes, films, slides, and colored lights. "Roaratorio" (1979) was an electronic piece comprising more than two thousand sounds mentioned in James Joyce's novel *Finnegans Wake*.

He was elected to the American Academy of Arts and Sciences in 1978. In 1982, the French Government awarded him its highest honor for exceptional contribution to cultural life, Commandeur de l'Ordre des Arts et des Lettres.

In this book's Introduction, I mentioned my interest in chaos and how small events in our lives can have amplified effects through time. John Cage wrote on this topic in an autobiographical statement: "When I wish as now to tell of critical incidents, persons, and events that have influenced my life and work, the true answer is all of the incidents were critical, all of the people influenced me, everything that happened and that is still happening influences me."[6]

One such influence was Cage's father—an inventor with far-ranging interests in electrical engineering, medicine, submarine travel, and even space travel. His father told John that when a person says something "can't be done" this simply pointed the direction to something that *should* be done.

According to Cage's "Autobiographical Statement," which first appeared in print in 1991, neither of his parents went to college. Cage went to Pomona College in California where he was surprised to see dozens of students in the library reading copies of the same book. Always the pioneer and wanting fresh ideas for himself and his teachers, Cage walked into the library stacks and read the first book written by an author whose name began with Z. He received the highest grade in his class. However, the experience left him with the impression that college was not being run effectively, and he dropped out shortly thereafter.

After his two years at Pomona, Cage decided to go to Europe, where he became interested in modern music and painting. When he wrote to his parents, telling them he wanted to return to America, his mother wrote back, "Don't be a fool. Stay in Europe as long as possible. Soak up as much beauty as you can. You'll probably never get there again."

Cage's mind, like most of the geniuses in this book, roamed beyond the fields in which he was most famous. His interest in Zen Buddhism, with its emphasis on the interdependence of all things, is said to have stimulated many of his experiments in music and mind. He attended D. T. Suzuki's classes on Zen at Columbia University, and had said that his Zen studies convinced him that music was a means to opening the minds of people who created, performed, or listened to the music. His music was meant to coax people to consider new possibilities and to think laterally.

In the 1950s, he studied lichen, mushrooms, and other wild edible plants and cofounded the New York Mycological Society. Cage once wrote of the time that he enrolled in a class on mushroom identification, even though he was an expert on the subject. The teacher had a Ph.D. and was the editor of a publication on mycology. One day, the teacher showed the students a mushroom, gave historical information on it, and declared it was the edible *Pluteus cervinus*. *Pluteus cervinus* gets its common name, "deer mushroom," from its dull brown color. Cage was certain that the fungus was not *Pluteus cervinus*, because the specimen had "gills" on the stem. In fact, the mushroom was probably a deadly member of the genus *Entoloma*.

Cage wondered what to do. Should he point out the teacher's error and thus make the teacher look bad in the eyes of the students? Or should he tell the teacher what the problem was, thereby possibly saving the lives of classmates. Cage eyed the students and decided to challenge the teacher, saying "I doubt whether that mushroom is *Pluteus cervinus*. I think it's an *Entoloma*."[7] The teacher responded by examining a key to mushroom identification, and realized that Cage was correct. The teacher walked over to Cage and said, "If you know so much about mushrooms, why do you take this class?" Cage responded that he still had much to learn. The teacher nodded—then silence ensued. Cage finally asked the teacher why the teacher did not identify the poisonous mushroom, and the teacher responded that he was a *specialist* in "jelly fungi" and that he usually didn't concern himself with the "fleshy fungi."

Isn't it fascinating the degree to which people specialize these days? A. J. Jacobs, author of *The Know-It-All*, was surprised to learn that many famous, smart people "before the 20th century held at least one second job that had absolutely nothing to do with the first," and he uncovered such unlikely job combinations as: poet/meteorologist, lawyer/astronomer, and lyricist/mollusk scientist. He notes: "Nowadays, not only do you have to specialize, but you have to specialize within your specialty. There probably aren't any general mollusk scientists anymore. You have to be a Northeastern digger clam reproductive scientist."[8]

Our society desperately needs generalists who traverse several fields and then bring together ideas in ways that specialists may be unable to do, but we seem to be learning more and more about less and less until we know everything about nothing. Sometimes, specialists develop blind spots after years of intense focus on a single topic. Even New

York City's specialized high school exam discourages generalization at an early age. According to the *New York Times*, a student with a 99 percentile score in math and 49 percentile in verbal would have been admitted to the prestigious Stuyvesant High School—but a student with a 97 in math and 92 in verbal would not.

A. J. Jacobs—famous for his reading of the entire *Encyclopaedia Britannica*—always wanted to be a generalist. In college, he "specialized" in introductory classes—taking all those "intro to" sociology, anthropology, math, and other classes. By his senior year, when his friends had all progressed to taking specialized seminars like "The Semiotics of Ornithology in Cervante's Oeuvre," he was still taking intro classes. He writes, "After college, I became a journalist partly because I could remain something of a generalist. That, and I had no other job offers."[9]

Cage was also a generalist and quite the Renaissance man, but his far-reaching interests pale in comparison to the geniuses of yester-year. Consider as just one example, Johann Wolfgang von Goethe (1749–1832)—lawyer, painter, statesman, soldier, poet, novelist, philosopher, botanist, biologist, color theorist, and mine inspector. Goethe was one of the most important European literary figures of his time. He was the author of *Faust* and *Theory of Colors* and inspired Darwin with his independent discovery of the human premaxilla (upper jaw) bones, now named "os Goethei" in his honor. Goethe studied rocks and minerals, and the mineral goethite was named after him.

Goethe's Bones

John Cage collected and studied mushrooms—Goethe fell in love with bones. In Goethe's day, comparative anatomists realized that humans and other mammals had virtually the same anatomical structures. One of the last major holdouts for differentiating humans from the "lesser beasts" was the premaxillary bone that forms the end of the snout of the upper mammalian jaw. According to anatomists of the day, humans lacked these bones, and thus separated from other "lower" animals.

Judeo-Christian sentiments led people to believe that key differences must exist between humans and animals. Sadly for the religious folks, Goethe believed that the concept of humanness could not hinge upon the premaxillary bones. His epic quest to find the premaxilla finally ended when his access to many skulls allowed him to carefully study the region behind our teeth from inside the mouth. There, Goethe saw in some skulls, two little premaxillary bones that were positioned *between* the upper jawbones. Others had made the same observations as Goethe, but their work had been ignored or forgotten. Goethe was so happy with his discovery that he wrote to friends,

> I have found—neither gold nor silver, but it gives me the greatest joy—the premaxilla of the human being. . . . It is like the keystone to the human being; it's not missing; it's there! The difference between the human being and animals is not to be found in any given particular. . . . The human being is a human being just as much through the form and nature of the upper jaw as through the form and nature of the last two bones of the little toe.[10]

Goethe is perhaps best known for his drama *Faust,* which was only published in its entirety after his death. Both Goethe and the character of Faust were intrigued by the notion of immortality and the existence of an immortal soul. Perhaps this is one reason why, in Milan Kundera's novel *Immortality,* Goethe and Ernest Hemingway meet in heaven and actually debate whether they or their books are what has brought them fame. "Instead of reading my books, they're writing books about me," Hemingway says. "That's immortality," Goethe replies. "Immortality means eternal trial." In real life, Goethe once remarked that "life is the childhood of our immortality."

Goethe was one among several eighteenth- and nineteenth-century polymaths still praised in the twenty-first century. For another, consider Samuel Johnson (1709–1784)–the scholar, poet, and playwright–who suffered from obsessive-compulsive disorder. He was a hypochondriac, had auditory hallucinations and facial tics, and was compelled to go through elaborate rituals while entering any doorway. Despite his handicap, his encyclopedic and wide range of interests included science and manufacturing processes, and his book *Journey to the Hebrides* was an important contribution to sociology and anthropology. Johnson was one of the most fascinating characters of the eighteenth century. He wrote satire and famous essays on Shakespeare and other poets. His monumental *Dictionary of the English Language* (1755) was the first general dictionary of English. He fancied himself a brilliant man, and he surely was. In fact, an entire "age" was named after him.[11]

Another chameleonic genius was Francis Galton (1822–1911), statistician, experimental psychologist, explorer, anthropologist, inventor, and eugenicist. A prolific author of over three hundred publications, Galton achieved his greatest fame with his book *Hereditary Genius* in which he argued that talent, and virtually every human trait, was usually inherited. He also conducted pioneering studies of human intelligence and statistics. He coined the words "anticyclone" in meteorology and "eugenics" in behavioral genetics. His research covered such diverse subjects as fingerprinting for personal identification, correlational calculus (a branch of applied statistics), blood transfusions, twins, criminality, and the art of exploration in undeveloped countries. Galton was a practical scientist always testing his theories using apparatuses of his own design. He was the first to capture and bring bizarre, blind amphibians (proteus) to England from the Postojna cave located in modern-day Slovenia and the first to publish a rigorous statistical study on the efficacy of prayer.

Galton had an obsessive predilection for quantifying anything that he viewed–from the curves of women's bodies to the number of brush strokes used to paint his portrait! In 1897, he published a paper in *Nature* on the length of rope necessary for breaking a criminal's neck without decapitating the head. In short, Galton was obsessed with the idea that anything could be counted, correlated, and understood as some sort of pattern.[12]

Silence

Galton cared about rigorous testing and measuring. Cage, on the other hand, was happy with chaos and delighted even if no content existed to organize and correlate. For example, the concept of emptiness in Cage's "4'33"" piece still confounds critics—and "4'33"" has had no direct precedent in history. It was Cage's attempt to express nothing while communicating everything through the vehicle of silence. Robert Lawrence made a Zenlike remark in *Closer to Truth*, "Sometimes only silence gets us closer to truth." Silence has always been a mystical topic in religion. The Buddha himself, when confronted with paradoxes, would reply with silence.

Even God responds with silence. Timothy Ferris in *The Whole Shebang* notes, "In a creative universe, God would betray no trace of his presence, since to do so would rob the creative forces of their independence, to turn them from the active pursuit of answers to mere supplication of God. And so it is: God's language is silence."[13] The Ismailis, members of a Moslem sect, developed a method of reading the Koran called *tawil* in which they attempted to hear a sound of a verse on several levels simultaneously, which helped them become conscious of the silence surrounding each word.

The concept of silence has played crucial roles in various entertainments. The title of Ingmar Bergman's 1963 movie *Silence* refers to the silence of God. A 1997 Iranian movie named *Silence* features a blind boy who earns a living tuning instruments. Another Iranian movie, inspired in part by the Taliban's rule in Afghanistan, is *The Silence between Two Thoughts*, which features a young woman waiting to be executed but then forced to marry the executioner. Both movies were banned in Iran.

The Twilight Zone episode named "Silence" features Archie Taylor who bets Jamie Tennyson $500,000 that Jamie can't stay quiet for a year. Jamie is able to remain silent, but is unable to collect his money because Archie has gone bankrupt. Sadly, in order to ensure winning the bet, Jamie had his voice box severed.

The silence in Thomas Harris's movie and book *The Silence of the Lambs* refers to the haunting childhood memories of FBI agent Clarice Starling who recalls the sounds of lambs being slaughtered. In the end of the movie, when she has solved some of the crimes, she is asked, "Well, Clarice, have the lambs stopped screaming?" Finally, in Simon and Garfunkel's song "The Sound of Silence," "silence" represents the lyricist's existential alienation from society, from the surrounding city, and from the very fabric of reality.

In the Hebrew Bible, have you ever wondered who is the last person with whom God speaks? The answer is Job, the human who dares to challenge God's moral authority. After the book of Job, God never speaks again, and He is decreasingly spoken of. In the Book of Esther—a book in which the Jews face a genocidal enemy—God is never mentioned. Why is there the long twilight in the Hebrew Bible where God is silent in the ten closing books? During the time of Moses, God is revealed through miracles. After the Jews are exiled to Babylon and return, there are few divine signs. John Fowles in *The Magus* sums up this isolation: "There are times when silence is a poem."[14]

❧ ❧

The rest is silence.

—dying words of Shakespeare's Hamlet

My favorite music is the music I haven't yet heard. I don't hear the music I write. I write in order to hear the music I haven't yet heard.

—John Cage, "An Autobiographical Statement"

At all the moments of death, one lives over again his past life with a rapidity inconceivable to others. This remembered life must also have a last moment, and this last moment its own last moment, and so on, and hence, dying is itself eternity, and hence, in accordance with the theory of limits, one may approach death but can never reach it.

—Arthur Schnitzler, *Flight Into Darkness*

Rather than complaining about the politics, I think that we should become actively disinterested in government. It seems to be the most active thing to do now.

—John Cage, "Interview with Laurie Anderson"

This reminds me that at dinner one night in 1966, John Cage showed me how he dealt with publicity: he weighed it on a small kitchen scale. Then he wrote down title, publication, date, and weight, and placed the press articles in a cabinet . . . unread.

—John Brockman, personal communication

The notes I handle no better than many pianists. But the pauses between the notes? Ah, that is where the art resides!

—Artur Schnabel (1882–1951)

"4'33''" is music reduced to nothing, and nothing raised to music. It cannot be heard, and is heard anywhere by anyone at any time. It is the extinction of thought, and has provoked more thought than any other music of the second half of the twentieth century.

—Paul Griffiths, *Modern Music and After: Directions Since 1945*, 1995

What is music? This question occupied my mind for hours last night before I fell asleep. The very existence of music is wonderful, I might even say miraculous. Its domain is between thought and phenomena. Like a twilight mediator, it hovers between spirit and matter, related to both, yet differing from each. It is spirit, but it is spirit subject to the measurement of time. It is matter but it is matter that can dispense with space.

–Heinrich Heine, *Der Salon*, 1836[15]

The function of music is to release us from the tyranny of conscious thought.

–Sir Thomas Beecham (1879–1961)[16]

Slugs can sense vibration. However, they are not capable of perceiving music as music. Mozart symphonies have no potentiality of existing in relation to the domain of knowledge of which a slug is capable. Although slugs and Mozart symphonies inhabit the same universe from our point of view, relative to the "point of view" of a slug, a Mozart symphony may be said to inhabit a different universe. Possibly it is better to say that both are part of the same "multiverse." The key point for reflection is that maybe we are like slugs.

–Charles L. Harper, Jr., "What does a Slug know of Mozart?" in *Spiritual Information*

THREE

GILGAMESH, GOD, AND THE LANGUAGE OF ANGELS

In which we encounter Maidanek death butterflies, crawling brains, brains in jars, the doll-people of the Popol Vuh, Elisabeth Kübler-Ross, cryonics, near-death experiences, Emanuel Swedenborg, glimpses of hell, Gehennah, Tophet, Jewish Sheol, the Witch of Endor, H. P. Lovecraft, devils and demons, Beelzebub, Iblis, ash-Shaytan, Jesus in Hell, *Gilgamesh,* Austen Henry Layard, Utnapishtim, Mount Mashu, George Smith, the Flood, the afterlife, Susan Blackmore, William James, *Varieties of Religious Experience,* John Nash, schizophrenia, the power of placebos, the quest for reality, over-belief, John Dee, John Kelley, Enochian records, Terence McKenna, Plato, and DMT-containing plants.

The greatest discovery of my generation is that a human being can alter his life by altering his attitudes of mind.

—William James

Disembodied Brains

The Brain from Planet Arous was one 1950s movie of several that involved disembodied, living brains. As medical technology advanced and "brain movies" proliferated, curious people wondered if brains in jars could be kept alive and conscious, and what impact this would have on an individual's afterlife. Philosophers even wondered what it would feel like to be a thinking brain, living in a vat.

The movie *Fiend Without a Face* (1958) featured human brains that crept along floors and tabletops and then leaped at victims. The brains' spinal cords trailed behind them like tails. If you buy the DVD, watch this movie with the lights turned low.

Fiend Without a Face takes place at an American Air Force base where unexplainable deaths occur near the base's atomic reactor. (Yet another 1950s movie reference to atomic power!) Autopsies performed on the bodies reveal that the brains and spinal columns are missing, somehow drained through two holes at the base of the skull. The hero, Major Cummings, soon discovers that disembodied brains lurk around every corner—the unfortunate result of "invisible thought beings" created by a local professor. The beings suck people's brains out and steal atomic power.

Another 1950s brain movie is *Donovan's Brain* (1953), directed by Felix Feist who gained fame in the 1960s for directing *The Outer Limits* TV show. The movie features Dr. Cory, a scientist who experiments with monkey brains as he tries to keep them alive in fish tanks filled with a cloudy fluid. Cory's wife is played by Nancy Davis, better known for her role as former First Lady Nancy Reagan.

The monkey brains in Dr. Cory's tanks are active and conscious as evidenced by brain waves traced on an oscilloscope. Could Dr. Cory keep a *human* brain alive in the same environment? As luck would have it, a nearby plane crash provides Dr. Cory with a terminal accident victim. Dr. Corey removes the brain and places it in a fluid-filled jar. The brain belongs to a Donald-Trump-like millionaire named Mr. Donovan. To make a long story short, the brain of this wealthy industrialist grows, glows, pulsates, and exerts a negative psychic force on those nearby. It communicates telepathically and causes Dr. Cory to become aggressive and carry out the brain's every wish.

As far out as *Donovan's Brain* was, it seems likely that a hundred years from now, if my body is destroyed in an accident but my brain is undamaged, there will be some way to remove the brain and keep it alive. Obviously, *Donovan's Brain* focused on a popular idea because the 1953 version is one of three movies based on the same story! The first was filmed in 1944 as *The Lady and the Monster* and last in 1963 as simply *The Brain.*

After watching these movies, it became obvious to me that the protagonist scientists really needed to widen their interests and get a life.

Their peculiar scientific obsessions always overwhelmed their social interactions. Decades earlier, fantasy and horror writer H. P. Lovecraft also featured men who were even further removed from normal society. His stories often revolved around lonely scientists and overdriven students who seemed to have few or no family members as they undertook their obsessed quests for knowledge. Even before Lovecraft's protagonists encounter dreaded mysteries, they can barely cope with everyday experiences. They come to realize that human life pales in comparison to the vast alien intelligences that rule the universe—cosmic fungi, ancient winged aliens, and eel-like creatures that float in a parallel universe. In the end, Lovecraft's human "heroes" sometimes go insane because they have used science to swim a little too deeply in the endless sea of mystery around us. In 1926, Lovecraft wrote in *The Call of Cthulhu*:

> The most merciful thing in the world, I think, is the inability of the human mind to correlate all its contents. We live on the placid island of ignorance in the midst of black seas of infinity, and it was not meant that we should voyage far. The sciences, each straining in its own direction have hitherto harmed us little; but some day the piecing together of dissociated knowledge will open up such terrifying vistas of reality, and our frightful position therein, that we shall either go mad from the revelation or flee from the deadly light into the peace and safety of a new dark age.

Immortal Brains

Although *Donovan's Brain* did not discuss *freezing* the brain, I wonder if the movie could have been made more interesting if scientists had cooled a brain sufficiently to make it enter a state of suspended animation, and then resurrected it centuries later. Here's a question for those who believe in an afterlife. If we were able to freeze *your* disembodied brain, and then revive you in a century, did you enter the afterlife during the intervening time of zero brain activity? If not, when *would* you enter the afterlife if your brain were frozen? If you did enter the afterlife during the frozen period, did you *return* from the afterlife when your brain was thawed?

My cryonicist friends believe that memories still reside in the brain cell interconnections and chemistry, and that the memories are preserved when a brain is cooled. Maybe they are right and technology in the future will be sufficiently advanced to revive people's brains that are frozen. After all, far back in the 1950s, hamster brains were partially frozen and revived by British researcher Audrey Smith.[1] If hamster brains can function after being frozen, why can't ours? In the 1960s, Japanese researcher Isamu Suda froze cat brains for a month and then thawed them. Some brain activity persisted.[2] In 1974, Suda showed that even after seven years of storage at –20° C, he could detect an abundance of seemingly normal cerebellar activity as well as short-lived EEG activity from the cerebral cortex.[3]

In the early 1980s, E. Haan and D. Bowen, of London's Institute of Neurology, froze sections of human cerebral cortex, sliced from living human patients. When the brain slices were thawed, they exhibited nearly normal metabolism.[4] In 2004, Peter Safar at the University of Pittsburgh replaced the blood of fourteen dogs with cold saline solution. After sixty minutes of being unconscious, with no breathing and no heartbeat, Safar was able to revive them by reinfusing blood. The dogs seemed to have no functional or neurological ill effects. In 2005, researchers at the Fred Hutchinson Cancer Research Center in Seattle slowed mice's metabolic rate to a near standstill by exposing the creatures to low concentrations of hydrogen sulfide gas—with no apparent ill effects. Respiration dropped from the normal 120 breaths per minute to fewer than ten, and body temperatures dropped from the normal 37° C to as low as 11° C.

In 2006, Hasan Alam, a trauma surgeon at Massachusetts General Hospital in Boston, placed pigs into a state of suspended animation—his cold pigs have no pulse, no blood, no electrical activity in their brains, and the pigs' tissues consume no oxygen. After a few hours, Alam pumps warm blood into the animals, the pigs' hearts start beating, and the animals jolt back to life.

According to the January 23, 2006, *New Scientist*, Alam has reanimated nearly two hundred pigs, with no perceivable negative effects. According to Alam, "Once the heart starts beating and the blood starts pumping, voilà you've got another animal that's come back from the other side. . . . Technically I think we can do it in humans." Of the reanimated mystery pigs, *Wired* magazine wrote, "Long the domain of transhumanist nut-jobs, cryogenic suspension may be just two years away from clinical trials on humans."

We will someday be able to use similar methods on humans, freezing them for a period of time, while they await lifesaving medicine or organ transplants. Perhaps the minds trapped in frozen brains will make a temporary excursion to the afterlife and return to tell us about the evanescent adventure in paradise.

Adventures in Hell

I was leafing through Jenna Jameson's autobiography *How to Make Love to a Porn Star* when I thought about certain religious traditions and the afterlife. Righteous Muslim men who die are rewarded with seventy-two virgins in heaven, but it is not clear to me that most men would want this. If you sought deep romantic love, wouldn't seventy-two be too many? And if you wanted wild sex, wouldn't you prefer seventy-two porn stars who could offer acts that were more creative, or better fake their delight than seventy-two virgins? Finally, the sexual act may be painful or frightening for a virgin woman. Would anyone delight in giving a woman pain?

A few scholars have suggested that a proper translation of the Koran and related Hadith (Islamic traditions) indicates that the Arabic word normally translated as virgin is actually a "white raisin" of crystal clarity. Whatever the case, I think if I had my choice, I'd prefer one woman at a time to truly be in love with in heaven, and, if that's not an option, then I would take the seventy-two porn stars. Naturally, my last choice would be seventy-two white raisins.

Some scholars have wondered from where the seventy-two virgins originate. Are they created on-the-fly from God, or are they taken from Earth? One friend suggested that the virgins may be beings like zombies who do not exhibit the full range of human emotions. Another colleague wondered if the virgins would somehow maintain a mystical virgin status, even after an infinite time with their dead lovers in paradise.

What if you died and then were returned from an afterlife in which you were not in a heavenly realm but rather in hell? Perhaps you've heard of near-death experiences (NDEs) in which people report pleasant visions when they are near death or "return" from death, for example when their heart has stopped and physicians were able to restart it. You've heard the stories about seeing a light at the end of a tunnel or seeing the smiling faces of loved ones—but there are also examples of near-death experiences involving hell! Many people have reported NDE trips involving tortures by elves, giants, and demons. Some parapsychologists consider good and bad NDE trips as evidence of heaven and hell.

Howard Storm had a particularly disturbing experience in the afterlife. Storm, who earned a master's degree from the University of California, Berkeley, was an arts professor at Northern Kentucky University. He was an atheist, and never believed in an afterlife—until 1985 when he died and went to hell. During his near-death experience in a Paris hospital while awaiting emergency surgery, he says he experienced the feeling of leaving his body. He spoke to his wife, but she could not hear him. Suddenly, he heard voices calling him from the hallway, which was filled with fog. He writes, "The people who were calling me were 15 or 20 feet ahead, and I couldn't see them clearly. They were more like silhouettes, or shapes, and as I moved toward them they backed off into the haze. . . . I knew that we had been traveling for miles, but I occasionally had the strange ability to look back and see the hospital room. My body was still there lying motionless on the bed."[5]

As Storm walked toward the voices, a thick, dark, fog obscured his surroundings, and he told the mystery beings that he did not trust them. They responded by screaming and biting him. He fought back, but there were too many of them. They began to strip his flesh from his bones. Suddenly remembering a prayer from a childhood Sunday school class, he said, "Yea, though I walk in the valley of the shadow of death, I will fear no evil, for thou art with me." The creatures backed away, cursing God. Storm screamed into the darkness, "Jesus, save me!"

Suddenly, a tall being materialized from a faraway pinpoint of light. A bright glow surrounded the being, and Howard felt ecstasy as his "body" was healed.

After Howard Storm recovered from his near-death experience, he entered United Theological Seminary and was ordained as a minister of the United Church of Christ. He documented his near-death experience in the book *My Descent into Death and the Message of Love Which Brought Me Back.*

Today, our near-death experiences are probably shaped by culture and popular media. For example, movies like *Constantine, Flatliners, The Cell,* and many others provide startling depictions of hell or hell-like worlds, as did popular media through the centuries—for example Dante's fourteenth-century *Divine Comedy.* To what degree are NDE visions of hell projections from the experiencer's mind?

Near-death researcher and author P. M. H. Atwater believes that the percentage of hell-like near-death experiences (NDEs) is probably much larger than has been previously claimed. She writes,

My first introduction to the NDE was in a hospital room listening to three somber people describe what they had seen while technically "dead." Each spoke of grayness and cold, and about naked, zombie-like beings just standing around staring at them. All three were profoundly disturbed by what they had witnessed. One man went so far as to accuse every religion on earth of lying about the existence of any supposed "heaven." The fear these people exhibited affected me deeply.[6]

Some of you may picture the devil as ruler of hell, inflicting physical and mental pain on others. But this portrayal of the devil is nowhere in the Bible. In the Bible, the devil is just another captive.

As I discuss in *The Paradox of God and the Science of Omniscience,* the Old Testament mentions Sheol, an unpleasant place in the afterlife from which the godly would be delivered. The New Testament mentions Hades and Gehennah. Hades, a word borrowed from Greek mythology, is a general term for the afterlife. Gehennah, on the other hand, was a place of everlasting torment. Back in the time of King Solomon, a temple in Jerusalem called Tophet existed where infants were burned. Tophet was in the Valley of Ben Hinnon, later called Gehennah in the New Testament. In this valley, children were sacrificed to the Canaanite god Moloch. The area was later used as a garbage pit, where fire burned almost continuously. Thus, Gehennah came to be a fiery metaphor for Hell.

The ancient Jews believed that Sheol or Tophet was a gloomy place of departed souls that were not tormented but wandered about unhappily. In the New Testament, Gehennah became a place of punishment.

The location of Sheol was generally thought to be underground, perhaps because of the long custom of burying the dead. In 1 Samuel 28:15, King Saul asks the Witch of Endor to summon the spirit of the prophet Samuel from Sheol. When Samuel materializes, he asks, "Why have you disturbed me and brought me *up*?"–clearly implying that Sheol and the realm of the dead are located beneath the earth's surface. The early Hebrews did not emphasize life after death, but during later difficult times of the Babylonian exile, the concept of an afterlife began to make sense to the Jews–if justice was not apparent on Earth, maybe it would be in an afterlife.

Jews and Christians refer to the devil as Satan, a fallen and arrogant angel. In parts of the Old Testament, Satan is not God's enemy but rather a challenger or accuser. The word "devil" comes from the Greek *diabolos*, meaning "slanderer," or "accuser." The word "Satan" is the English transliteration of a Hebrew word for "adversary" in the Old Testament, where he gambles with God about the faith of Job. Later, in the New Testament, Satan becomes the "prince of devils" and has names such as Lucifer (the fallen angel of Light), Belial (lawless), or Beelzebub (Lord of Flies):[7]

> All the people were astonished and said, "Could this be the Son of David?" But when the Pharisees heard this, they said, "It is only by Beelzebub, the prince of demons, that this fellow drives out demons." (Matthew 12: 24–27)

> What harmony is there between Christ and Belial? What does a believer have in common with an unbeliever? (2 Corinthians 6: 15–16)

For Christians, Satan's job is to tempt man to commit immoral acts. Moslems believe in Iblis, the personal name for the devil. They also call him ash-Shaytan, which means the demon. In the Koran, God tells Iblis to bow in front of Adam, the first human. Iblis refuses.

Seven Old Testament books and every New Testament writer refer to Satan. In the Middle Ages, theologians debated about how a supernatural being like Satan could exist in a universe governed by an omniscient, omnibenevolent, omnipotent God. Many came to believe that Satan was not an actual being but a symbol of evil.

According to Atwater, the researcher of near-death experiences, the word "hell" is Scandinavian in origin and refers to Hel, the Teutonic queen of the dead and ruler of "the other world." Legends indicate that fairly decent people went to Hel, if they were not sufficiently good and heroic to get into Valhalla. To the ancient Scandinavian people, Hel was not such a bad place in which to reside; however, Hel, the Goddess, was said to be deformed, with half of her face human and the other half featureless.

When reviewing hell-like near-death encounters, Atwater never encountered an individual who reported a fiery or burning sensation during the experience. More often she found descriptions of waning light, grayness, and fog. "Invariably an attack of some kind would take place in hellish scenarios," she writes. None of the children with NDEs saw visions of hell. Such visions are restricted to adults.

Christians who adhere to the "Apostles' Creed" are probably aware that Jesus went straight to hell after he was crucified. The Creed dates from very early times in the Church, about a half century after the last writings of the New Testament. The earliest written version of the creed is perhaps the *Interrogatory Creed of Hippolytus* (ca AD 215). It starts:

> I believe in God, the Father Almighty, the Creator of heaven and earth, and in Jesus Christ, His only Son, our Lord: Who was conceived of the Holy Spirit, born of the Virgin Mary, suffered under Pontius Pilate, was crucified, died, and was buried. He descended into hell.

According to early believers, Jesus invaded hell right after his death and before his resurrection. As told in the Gospel of Nicodemus, which was "accepted on par with the other Gospels until the New Testament was canonized,"[8] Satan actually imprisoned Jesus in hell. Jesus finally battles Satan and frees everyone trapped in hell, including famous Biblical figures like Adam, Eve, and Moses. Similarly, in Peter 3:19, we find that Jesus "went and preached unto the spirits in prison" after his crucifixion, which some have also taken to mean that Jesus went into Hell. However, some of my Roman Catholic friends were taught that the "Hell" mentioned in the Creed was not the full Hell but merely Purgatory in which souls were trapped and denied the presence of God until they atoned for their sins.

Gilgamesh and the Quest for Transcendence

All of my explorations regarding hell, heaven, and the afterlife eventually led me to Nancy K. Sandars's 1972 translation of the *Epic of Gilgamesh*. *Gilgamesh* is humankind's oldest recorded tale of a hero and a thousand years older than the *Iliad* or the Bible. The original was written in the Sumerian language and recorded in cuneiform characters on clay tablets that were found at Nippúr in Mesopotamia. The tablets date back to around 2000 BC and contained mysterious wedged-shaped symbols resembling the shapes shown here:

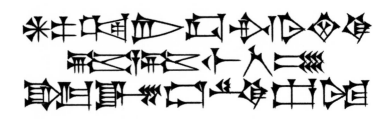

Today, the poem survives only in pieces on thousands of small tablets that were discovered scattered across a wide area now in modern Iraq. We don't know who wrote the epic. The tale was crafted and reworked by various Mesopotamian cultures including the Sumerians, Akkadians, Babylonians, and Assyrians. The names in the story are as mysterious as all the gods, monsters, and customs. The original story appears to take place around 2700 BC during the time when a great king ruled the city-state of Uruk.

The *Epic of Gilgamesh* is one of my deepest obsessions. To me, the *Epic of Gilgamesh* is a portal to a parallel universe. It gives its readers a glimpse of other ways of thinking and of other worlds. It is also among the most mysterious books ever written. Like the Bible, we don't know the ratio of myth to history. We are not always sure of the intended message. We only know that the *Epic of Gilgamesh* reflects humankind's deepest feelings as it chips away at the edges of our unconscious. At a minimum,

house of dust and I saw the kings of the earth, their crowns put away forever; rulers and princes, all those who once wore kingly crowns and ruled the world in the days of old.[9]

Our hero Gilgamesh is depressed by this vision of the afterlife and begins his quest for immortality. But who can help answer his deepest questions? He is familiar with the legend of Utnapishtim, a mortal man whom the gods made immortal, and thus Gilgamesh decides he must meet this legendary man.

The scenes of Gilgamesh's journey to the edge of the world–to lands beyond death and to realms of space close to the gods–continue to resonate in my mind and haunt me. Utnapishtim lives at the mouth of all rivers, at the end of the world. After a long journey, Gilgamesh arrives at Mount Mashu, which guards the rising and the setting of the sun and "whose flank reaches as far as the Netherworld below." Two large scorpion beings protect the way. Gilgamesh shields his face, because just a glance of their faces means death. They tell him that no man has ever journeyed this far, but they do allow him to pass into "the land of Night," which has never seen the light of day.

The scorpion people gesture to a tunnel into which the sun sets each day and through which it returns the next day. Gilgamesh must enter the tunnel and traverse its length before the sun descends into the passage and burns him to a crisp.

After a long journey running through the dark tunnel or passageway, Gilgamesh emerges in a sparkling garden of jewels, where every tree bears precious gems and stones. In this heavenly realm, Gilgamesh gazes with awe on glistening plants made of carnelian, towering lapis lazuli trees, and bushes made of jewels and coral. Stephen Mitchell, author of *Gilgamesh: A New English Version*, calls this journey "a symbolic death and rebirth, in which he passes through the darkness of an underworld and emerges into the dazzling, *Arabian Nights*–like garden of the gods."

After additional travel, Gilgamesh comes to another dark place where he must cross the Waters of Death on a boat. Any mortal who touches the waters dies instantly, but a strange boatman, along with his crew of Stone Men, are the only beings who can help Gilgamesh travel across and survive. Apparently, the Stone Men can wade in the Waters of Death and survive. For reasons unclear to us today, Gilgamesh smashes the Stone Men. Luckily, the boatman says they can use long poles to help propel the boat instead.

The portion of Gilgamesh that describes the legend of Utnapishtim, the Flood survivor, is similar to the Hebrew account of Noah and the Flood. In fact, the Victorian discoverer of the Gilgamesh flood, George Smith (1840–1876), was dumbfounded when he realized that the Bible's Noah story appeared much earlier in *Gilgamesh*. In his shock, he is said to have torn off his clothes, surprising his colleagues at the British Museum.[10] Other Old Testament stories, such as the Tree of Life and Samson and Delilah, also may have their roots in the ancient *Gilgamesh*.

Let's review the events leading up to the discovery of the flood text in *Gilgamesh*. Smith had a day job as a bank-note engraver. At night, he studied cuneiform tablets, which had been transported to Egypt from the Middle East. After he published several insightful observations, the British Museum appointed Smith to an assistant position.[11] As he worked, trying to translate fragments of cuneiform tablets that had been excavated from the ancient Assyrian capital of Nineveh, he noticed a line which made him gasp: "an old man, Utnapishtim, took his family and all kinds of animals aboard the great boat, and the flood. . . ." The next piece was missing! The Assyrian texts predated the Bible, and he thought the inscription could only mean one thing: *proof* of Noah and the Flood.

"On looking down the third column," Smith wrote, "my eye caught the statement that the ship rested on the mountains of Nizir, followed by the account of the sending forth of the dove, and its finding no resting-place and returning. I saw at once that I had here discovered a portion at least of the Chaldean account of the Deluge."[12] Smith continued to gaze at the cuneiform writing. According to Stephen Mitchell, Smith shouted, "I am the first man to read that after more than two thousand years of oblivion!" Setting the tablet on the table, Smith jumped up and "to the astonishment of those present, began to undress himself."[13] I consulted a number of *Gilgamesh* experts, and none could tell me if Smith simply removed his jacket or he continued to remove all of his clothes in a state of temporary insanity.

Perhaps the Biblical story of Noah did come from *Gilgamesh*, or perhaps both versions derive from an even older source. Some anthropologists speculate that the story reflects a folk memory of events in 5000 BC, when melting glaciers caused the Mediterranean to overflow, flooding a large, densely settled area around the Black Sea and scattering its survivors.

News of 1872 discovery of the "Mesopotamian Noah" shocked London society, and the *Gilgamesh* story spread. However, the flood story that Smith discovered was incomplete, and people wanted to know how the tale ended. According to one account, a London newspaper offered a large reward to anyone who could find clay tablets that supplied the conclusion to the flood story.[14] More precisely, the *London Daily Telegraph* offered to fund an exhibition to search for the missing fragments. Smith accepted their 1,000-guinea offer and applied it toward the cost of the excavations for the missing Noah story. Smith raced to Nineveh where the original fragments had been found. In a flash of serendipity, he quickly uncovered the missing tablets among thousands in the ruins. In particular, Smith recovered from Kuyûnjik a fragment that contained the greater portion of seventeen lines of inscription that complete the story.

During his journey, Gilgamesh meets the Babylonian Noah, Utnapishtim. The reader learns that the Mesopotamian gods were angry with people and destroyed all of humanity in a great flood. One god, however, warned Utnapishtim about the impending flood—and Utnapishtim built an ark and sealed it with pitch to save his family. In

Utnapishtim's account, the continuous rain caused the water level of the world to rise to the roofs of the houses, and after the waters receded, Utnapishtim released a dove, a sparrow, and a crow—a very similar kind of story appears in the Old Testament. The destroyer for the Hebrews is Yahweh, and for the Mesopotamians it is Entil, one of many gods of the time. Similar in spirit to the Old Testament, the gods in *Gilgamesh* repented of their destruction of humanity because they needed humans for sustenance. One interesting difference from the Noah story revolves around Utnapishtim's decision to take more than animals aboard the ark to ensure the replenishment of civilization. He also took "craftsman and artisans of every kind" aboard his ark. Good move, but a crowded boat.

Utnapishtim and his wife are granted eternal life, and when Utnapishtim meets Gilgamesh, he gives Gilgamesh a plant that Gilgamesh can eat to live forever. Alas, a serpent eventually eats the plant, and Gilgamesh remains mortal. Gilgamesh ends his tale as part of the natural order and must accept death as part of life.

During the excavations that Smith carried out in 1873 and 1874, he recovered additional material from *Gilgamesh* that enabled him to complete his translation of the story of the Flood. Unfortunately Smith died of hunger, exposure, and sickness in 1876. He was only 36.

Elisabeth Kübler-Ross and Maidanek Death Butterflies

Gilgamesh's journey, past Mount Mashu through a long dark tunnel to a world beyond our ordinary Earth, reinforces the idea that our ancient ancestors were as obsessed as we are with death, dying, and the concept of the afterlife. Millennia later, serious scientists explore the twilight realm between life and death, and continue their own metaphysical quests in the tradition of Gilgamesh. One of the most famous modern-day explorers of death was Elisabeth Kübler-Ross, the psychiatrist whose pioneering work with terminally ill patients helped to revolutionize attitudes toward the care of the dying. She also spent part of her career on research to verify the existence of life after death. After conducting thousands of interviews with people who had had near-death experiences, she believed she had acquired sufficient evidence of an afterlife. However, some scientists were skeptical of her work and suggested she was possibly motivated by her own fears of her death.

Elisabeth Kübler-Ross (1926–2004) was one of three triplets and weighed only two pounds at birth. When she was in sixth grade, she knew she wanted to be a physician. However, her father scoffed at this plan, telling her she had no choice but to be a secretary in his office supply business. Like Capote and Taylor, her later success was not a result of parental encouragement.

Elisabeth had more than an ordinary share of death in childhood. When she was a young girl hospitalized for pneumonia, she shared a hospital room with a girl who eventually died in the room. A few years later, she visited a dying friend who was in a room filled with flowers and visitors, and with a nice sunny view. To her, this was a "good death."

Her early interest in death and dying gelled when she visited World War II concentration camps after the war in 1945. In Maidanek, a Polish death camp, she discovered butterflies that prisoners etched into the walls before they died:

> It started in Maidanek, in a concentration camp, where I tried to see how children had gone into the gas chambers after having lost their families, their homes, their schools and everything. The walls in the camp were filled with pictures of butterflies, drawn by these children. It was incomprehensible to me. Thousands of children going into the gas chamber, and this is the message they leave behind—a butterfly. That really the beginning.[15]

The butterfly became Kübler-Ross's symbol of a beautiful transformation that she believed occurred at the time of death. In 1962, she became a teaching fellow at the University of Colorado School of Medicine in Denver. One of her lectures had a profound impact on her and her class when she invited a teenage girl who was dying of leukemia. The students in the audience were asked to interview the girl. Naturally, they asked about the girl's blood tests and other related medical questions. The girl told them they were insane. Why weren't they asking her questions about what it *felt* like, never to dream of going to the high-school prom, or even growing up? Why didn't doctors ever tell her the truth?

At the end of the lecture, students cried in their seats. Dr. Kübler-Ross nodded, telling the students that they now knew how to act as human beings and not just clinical scientists. From this point on, her lectures were filled with standing-room-only crowds of medical and theology students, but few physicians.

Kübler-Ross began to notice that dying patients usually followed five stages of dying: denial, anger, bargaining, depression, and finally acceptance. She also found that most of the dying could accept death most easily if they could look back and feel that they had not wasted their lives. Stop reading for a moment and ponder this: Do *you* feel that you are wasting your life? Why?

In the late '60s, Kübler-Ross was fascinated by a woman who was pronounced dead in a hospital but later was revived and able to recount conversations of others in the room at the precise time she was clinically dead. The more Kübler-Ross contemplated this case, the more she wondered if life, or at least some kind of existence, were possible after death. She interviewed hundreds of patients who had come back from clinical death and who claimed that during the experience, they were free of pain and seemed to float out of their bodies. They felt as if they were being guided to a place of light and love and did not want to return. After hearing these experiences, Kübler-Ross concludes that we don't die alone and that a spirit being comes to comfort those who are about to die.[16]

Kübler-Ross's most controversial book was her 1992 *On Life After Death*, based on interviews with more than twenty thousand people, many of whom gave similar accounts of near-death experiences. In her book, she said that her job was "to tell people that death does not exist." Her 1997 autobiography listed four phases of the afterlife: leaving the physical body, meeting angels or spirit guides, traveling through a tunnel or bridge toward a light, and becoming pure spirit. The tunnel travel has deep resonances with the *Epic of Gilgamesh*, in which Gilgamesh travels through a dark passage and emerges in a brilliant garden of jeweled trees.

Most skeptics have no trouble with the notion that people with near-death experiences often see a light at the end of a tunnel. They can accept many of the unusual anecdotes in Kübler-Ross's book. For example, a Dutch study, reported in the prestigious *Lancet*, found that about 10 percent of 344 patients resuscitated after suffering cardiac arrest had NDEs, and about 18 percent remembered some part of what happened when they were clinically dead.[17]

As I report on in my book *Mazes for the Mind*, one of the most unusual applications of computer software in the 1990s has been to the study of visual experiences reported by people who have almost died. Many people, who have "come back" from states close to death, have reported pleasant experiences at death's door. Susan Blackmore, Visiting Lecturer at the University of the West of England, Bristol, along with colleague Tom Troschianko, used a computer program to answer the question: Why do the nearly dead see lighted tunnels?

Even before Blackmore and Troschianko, researchers in the past had shown that several patterns are likely to appear to people whose brains have been subject to drugs or abnormal electrical stimulation as occurs in epilepsy. These patterns include the tunnel, the spiral, the lattice or grating, and the cobweb. Blackmore has written in "Near-Death Experiences: In or Out of the Body" that she believes these patterns arise in the structure of the visual cortex, the part of the brain concerned with vision. Blackmore and Troscianko's computer program simulates what would happen when increasing electrical noise arises in the visual cortex. The computer program starts with thinly spread dots of light, with more toward the middle and very few at the edges of the pattern. (Blackmore notes that in the cortex there are many more cells representing the center of the visual field but very few for the edges.) When the computer simulation is run, gradually the number of dots increases, and the center begins to look like a white blob. The researchers were shocked to see on their display a dark speckled tunnel with a white light at the end. The light grows bigger and bigger (giving the appearance that the observer is getting nearer and nearer) until it fills the whole screen. Is this the tunnel some see at the threshold of death? It may be too early to answer this with any certainty. Blackmore notes, "Our program and theory also make a prediction about near-death experiences in the blind. If they are blind because of problems in the eye but have a normal cortex, then they too should see tunnels." If you wish to gaze at the eerie and crepuscular death-tunnels produced by their computer simulations, see Blackmore's 1991 *Skeptical Inquirer* article.[18]

Although scientists accept the NDE as a real experience and give credence to some of Kübler-Ross's studies, some aspects of her thinking were thought to be eccentric. Jonathan Rosen, writing for the *New York Times Magazine*, found it creepy when Kübler-Ross showed him a "future map" of the United States, dated 1998 to 2001, that showed huge areas of the country that would be decimated by environmental and other upheavals.[19] Large chunks of the US would be literally swept away. When Rosen asked her if this decimation scared her personally, she replied that it did not because no one really dies. He writes, "Kübler-Ross made a lasting contribution to the care the dying receive, but she is also a haunting reminder that battling death can darken the spirit in strange and unsettling ways."[20]

When her own health started to deteriorate around the year 2000, she said she was in pain and ready to make the transition. She spoke of her imminent death, "I am like a plane that has left the gate and not taken off. I would rather go back to the gate or fly away." Once death came, she predicted that she would be "dancing in all the galaxies." She died in 2004 at the age of seventy-eight after a series of strokes.

William James, John Nash, and Life After Death

Harvard professor William James (1842–1910), one of the principal founders of modern American psychology, also believed that spirits existed beyond our ordinary reality. He was the author of the popular *Varieties of Religious Experience* and the influential textbook *Principles of Psychology*, which contained elements of his personal philosophy and even autobiographical musings. James became the chief American advocate of "pragmatism," a philosophical theory that posits "that is true which works."

James was born in New York City to an intellectual, religious, and affluent family. His father made sure his children were well educated and frequented museums. One son, Henry, became one of America's most cherished novelists. William became one of the world's foremost explorers of the mind.

By the end of his teenage years, William James was fluent in five languages and had visited nearly every major museum in America. He initially aspired to become a painter, but his father urged a field in which James could better excel.

James began medical school at the age of nineteen—during the American Civil War. He suffered from bouts of depression—an affliction that appears to occur with a significantly higher percentage in certain fields of creative genius than in the standard population. He also suffered from panic attacks and occasional hallucinations. I discuss the association between mental illness, creativity, and genius in my book *Strange Brains and Genius*, but one more recent example is the Nobel Prize–winning mathematician John Nash, who has suffered from schizophrenia. He observes,

> There's certainly a connection between mental illness and "thinking out of the box." If you're going to be anything like a genius, you have to think out of the box. . . . Mathematicians are comparatively sane as a group; it's the people who study logic that are not so sane. Logical scholars like Kurt Gödel [deliberately

starved himself to death in 1978] are certainly not a good example of sanity.[21]

James vowed to use every ounce of his mental will to forge ahead to overcome his mental illnesses. Although he believed that predisposition to mental illness had a hereditary component—in fact several of his siblings were invalids at one time or another—he also believed that through logic and perseverance, he could alter his psychological states. He wrote, "My first act of free will shall be to believe in free will."[22]

I believe that the root of most severe mental illnesses are biological in origin, although cases appear to exist in which a person's coping techniques and logical assessment can help ameliorate some symptoms of mental disease. John Nash describes his approach to his schizophrenia and the voices in his head:

> Ultimately, I realized I am generating these voices in my own mind: this is dreaming, this is not communication. This is coming from an internal source, not from the cosmos. And simply to understand that is to escape from the thing in principle. After understanding that, the voices died out. My son hears voices, but I haven't heard any for a long time. . . .
>
> I think there is an element of choice. A person doesn't pass into insanity when their situations are good . . . if their personal life is successful. . . . Wealthy people are less likely to become schizophrenic than people who are not wealthy.[23]

I don't fully agree with Nash, because there are plenty of cases in which a person *does* pass into insanity even when their life situation is apparently good. But it is interesting that Nash should think this way. If I recall correctly, Nash himself was at the high point of his career and life when he started succumbing to his schizophrenia!

Evidently, William James was pretty successful in his drive to succeed and cope with his depression. After he had graduated with a medical degree from Harvard University at age twenty-seven, he was soon teaching philosophy, psychology, and physiology at Harvard. Students said that he was a wonderful and warm teacher, and that he taught them what it really means to be engaged in life and in one's passions. He conducted his initial studies of the mind in a small room next to the Harvard laboratory, where he collected sheep's heads and frogs.

When he became engaged to Alice Gibbens, he warned her about his bouts of deep depression and occasional suicidal thoughts. She married him anyway in 1878.

The chameleonic James was not content to study or work in a single field. He may have started as a brilliant teacher of physiology and then psychology, but in 1879 he switched to philosophy. In 1902, he wrote: "I originally studied medicine in order to be a physiologist, but I drifted into psychology and philosophy from a sort of fatality. I never had any philosophic instruction, the first lecture on psychology I ever heard being the first I ever gave."[24]

Despite his serious and traditional education, he surprised the scientific community by announcing that he was in contact with the spirit of his colleague and friend Dr. Richard Hodgson (1855–1905), who had died suddenly of heart failure at the Boat Club in Boston. James's 100-page report was published in the "Proceedings of the American Society for Psychical Research," which contained verbatim records of his ghostly conversation. Hodgson had believed in spirits and was one of the founding members of the American Anthropological Association.

James thought deeply about the philosophy of religion and believed that religion and religious studies should focus on religious *genius* instead of religious *institutions*, since the institutions were vestiges or vague shadows of the genius. According to James, psychologists should also seek religious experience as a means to understand reality and the mind. He urged us all to "over-believe," that is believe things that cannot be scientifically proven, because these over-beliefs help us live fuller and happier lives. He writes in *The Varieties of Religious Experience*:

The over-belief, on which I am ready to make my personal venture, is that [divine miracles] exist. The world of our present consciousness is only one out of many worlds of consciousness that exist, and that those other worlds must contain experiences which have a meaning for our life also; and that although in the main their experiences and those of this world keep discrete, yet the two become continuous at certain points, and higher energies filter in.[25]

As mentioned, his theory of pragmatism tends to suggest that we might as well define "true" as meaning "useful." He also believed that purely mental events and experiences had a validity on par with external reality. Most of my colleagues do not agree with James's pragmatism, but many suggest that something doesn't have to be true in the traditional sense in order to be useful. Many models in physics, for example Newtonian physics, may not be "true," but are treated as if they are true because they yield reasonably consistent predictions. Perhaps placebo effects in medicine, slightly inflated views of self-worth, and belief in God and the afterlife, can alleviate pain and even increase our life spans, if we do our best to treat these beliefs and ideas as true. Alas, it is difficult for many people to achieve the beautiful results of erroneous belief because it is difficult to believe something is true when one knows with relative certainty that it is not true. In the future, when scientists better understand the brain, it may be possible to compartmentalize the mind in order to allow us to derive the health benefits and comforts of erroneous belief while still permitting our minds to understand that the belief is not logically or scientifically true.

William James believed in the afterlife, and he publicly defended his support of personal immortality in his *Ingersoll Lecture on Human Immortality*. Even if the afterlife does not exist, various modern studies indicate that near-death experiences have significant value. Various studies conducted by Dutch cardiologist Pim Van Lommel and American pediatrician Melvin Morse suggest that even years after subjects' near-death experiences, the subjects had a decreased fear of death, an increased belief in the afterlife, a feeling that their lives had meaning, and increased social awareness and religious feelings. Compared to control groups, these people gave more money to charity, were more empathetic, took fewer medications, and were positively transformed by the experience. Even if the subjects' beliefs about the "meaning" of the near-death experiences were erroneous, they often derived clear benefit from them. Whether their thoughts and memories of the experience were part of this reality, and the result of neurochemical changes in the brain, or came from a realm beyond our ordinary existence, people often felt a transcendent and unifying feeling of love.

James believed that the brain served as a kind of filter for thought, which already existed in some other realm or reality. He writes, "We need only suppose that continuity of our consciousness with a mother-sea, to allow for exceptional waves occasionally pouring over the dam"—a dam that the brain creates in each person.[26] Knowledge and insight are not produced but rather "they exist ready-made in the transcendental world, and all that is needed is an abnormal lowering of the brain-threshold to let them through."

Emanuel Swedenborg

According to Reinhold Niebuhr, who wrote the introduction to Simon & Schuster's edition of James's *Varieties of Religious Experience*, James inherited his lifelong interest in religion from his Swedenborgian father. In fact, James was home-schooled by his father who tutored James in the writings of Emanuel Swedenborg (1688–1722)—Swedish scientist, mystic, and philosopher. Although James rarely gave direct references to Swedenborg in his writings, both James and his father felt that religion was a central, positive aspect of human life.

As I discuss in *Dreaming the Future*, Swedenborg's claims of communicating with the dead and with higher planes of being had a strong influence on modern occultism, spiritualism, and psychology. Although some of Swedenborg's contemporaries thought he was a nut case, after his death a religion called the Church of New Jerusalem was formed around his beliefs. His work even inspired a number of important people from Abraham Lincoln and Carl Jung to Helen Keller.

Swedenborg wrote very little about his early life, except for the following in a letter to a friend that shows the depth of Swedenborg's interest in God and the realm of spirit: "From my fourth to my tenth year, I was constantly engaged in thought upon God, salvation, and the *spiritual sufferings* of men. . . . From my sixth to my twelfth year my delight was to discourse with clergymen concerning Faith–that the life thereof is love, and the love that gives life is the love of one's neighbor. . . ."[27]

Swedenborg believed that an angelic language exists, a kind of protolanguage for beings residing on another plane of reality. You and I will speak this language in our afterlives:

> There is a universal language, proper to all angels and spirits, which has nothing in common with any language spoken in the world. Every man, after death, uses this language, for it is implanted in every one from creation; and therefore throughout the whole spiritual world all can understand one another. I have frequently heard this language and, having compared it with languages in the world, have found that it has not the slightest resemblance to any of them; it differs from them in this fundamental respect, that every letter of every word has a particular meaning. [28]

At the University of Uppsala, the chameleonic Swedenborg studied philosophy, mathematics, science, Latin, Greek, and Hebrew. After graduating, he studied physics, astronomy, and other natural sciences, as well as learning bookbinding, cabinet making, engraving, brass instrument making, and lens grinding. In just a few years, he absorbed a fantastic amount of knowledge and wrote extensively. He also made numerous original discoveries in a wide variety of scientific disciplines. For example, Swedenborg recognized the function of the brain's pituitary gland and developed plans for a glider airplane and a submarine. He presented mechanical-mathematical analyses of submarines, mechanical carriages, lock systems for raising ships, and "a method of conjuring the wills and affections of men's minds by means of analysis." In the area of chemistry, Swedenborg presented an atomic-like theory for matter: "The first conception is that all things are in series; there is a series of particles, beginning with mathematical points and ending in water, salt and earth."[29]

As he sought to better understand the universe and nature, he delved into anatomy and cosmology, the latter a field in which he was one of the first individuals to advance the nebular hypothesis by which our solar system was formed from a rapidly rotating cloud of gas. Throughout the period of his scientific work, Swedenborg maintained his interest in spirituality and the infinite. The aim of much of his biology research was to find a rational explanation for the operation of the soul.[30]

In 1743, Swedenborg said that God had given him a mission to interpret the Bible from a spiritual viewpoint. To help him, God allowed him to witness the afterlife where he studied the social organization of heaven, hell, and the realm where spirits go at the moment of death.

In his book *Heaven and Hell*[31], Swedenborg said that there was no devil, no single ruler of hell. People in hell might have the power to influence living people. Hell exists because it is the only place in which evil people are comfortable. In heaven, the evil people would suffocate from the love and goodness. According to Swedenborg, the creation of Hell was not an act of punishment but of divine mercy. When one dies, a person lives forever in the place that best matches the love the person fostered in his heart while on earth.

Swedenborg had countless theories about this world and the world beyond. For example, he said that everything in the physical world had a corresponding element in the spiritual world. He wrote that the Second Coming of Christ would not take place in physical form.[32] He believed that he had been called by God to give a new revelation to humanity, and for the next twenty-seven years, until his death in London at the age of eighty-four, he wrote thrity volumes of theological works that comprise that revelation. In the last month of his life, several of his friends asked Swedenborg to comment on his writing. He replied: "I have written nothing but the truth, as you will have more and more confirmed all the days of your life, provided you keep close to the Lord and faithfully serve Him alone by shunning evils as sins against Him and diligently searching His Word, which from beginning to end bears incontestable witness to the truth of the doctrines I have delivered to the world."

Swedenborg's followers, known as Swedenborgians, believe that his religious writings are divinely inspired. He never intended to found a new religious denomination, but in 1787 British printer Robert Hindmarsh organized his disciples in England as a separate sect. Today, Swedenborgian congregations exist throughout the world, with about fifty thousand members, and Swedenborg's theological writings have been translated into a large number of languages. The Swedenborgian General Church of New Jerusalem with headquarters near Philadelphia has about five thousand members in thirty-three churches. The Swedenborgian Church of North America, with headquarters near Boston, has about thirty-seven active churches with about fifteen hundred US members.

Swedenborg is not the only mystic to posit a universal language, spoken by the angels or by people in the afterlife. When British mystic John Kelley (1555–1597) gazed into the crystal ball, he claimed that "in the middle of the stone seemeth to stand a little round thing like a spark of fire, and it increaseth, and it seemeth to be as a glove of twenty inches diameter, or there about."[33] Through this glowing central sphere, Kelley claimed to be in contact with spiritual beings who had a message for his partner, John Dee (1527–1608), the noted British mathematician, astronomer, astrologer, geographer, and consultant to Queen Elizabeth I.

In particular, Kelley said that the beings wanted to teach Dee "Enochian," the language spoken by angels and the inhabitants of the Garden of Eden. Dee's unusual Enochian records are sufficiently detailed that some people were convinced they represented a genuine pre-Hebraic language. However, other researchers have suggested that Enochian was a code Dee used to transmit messages from overseas to Queen Elizabeth.

Terence McKenna, in his book *The Archaic Revival,* writes that Dee and Kelly "recorded hundreds of spirit conversations, including . . . an angelic language called Enochian, composed of non-English letters, but which computer analysis has recently shown to have a curious grammatical relationship to English."[34] The term "Enochian" comes from the prophet Enoch, the Biblical figure who "walked with God" and was taken directly to heaven, avoiding death. This script contains twenty-one letters, and each letter has its own name and pronunciation. Some of the characters, like Ω and λ, bear a striking resemblance to Greek letters. An example of Enochian, which the angels said had been spoken by the Old Testament prophet, Enoch, sounds like this: *Madariatza da perifa Lil cabisa micaolazoda saanire caosago of fifa balzodizodarassa iada,* which means, "Oh, you heavenly denizens of the first air, you are mighty in the parts of the earth, and execute judgment of the highest."

Here is a sampling of the Enochian characters[35] that I have used to create a secret message. So far no one has been able to decipher my secret code.

No one knows for sure from where the Enochian language comes. It does not appear to be from the ancient Middle East, although a few of its characters, like ٦ or ٦, look Hebraic. Supposedly, the spirits provided Dee and Kelly with predictions of future events. For example, Uriel, the most regular angelic visitor, explained that Mary Queen of Scotts was fated to be executed by Queen Elizabeth. Many "believers" have asserted that the Enochian language predates all human languages and could be used to contact intelligences from other dimensions. Dee and Kelley said they thought the language could be used to converse with the Nephilim, the giant human-angel hybrids of the Old Testament.

Dee and Kelley were not the first people to present some kind of "Enochian" alphabet. For example, an alchemy text called the *Voarchad-umia* by Pantheus, written in 1530, contained an eighteen-character alphabet also attributed to Enoch. Interestingly, the British Museum's copy of this manuscript has copious marginal notes by Dr. John Dee. The notes date from 1559, and although Dee's twenty-one-character alphabet does not resemble Pantheus's alphabet, it is possible that Dee got the idea from that text.

The Doll People of the Popol Vuh

Enoch never died, but, alas, Gilgamesh suffered his mortal fate like the rest of us will. I hope that someday more people will read the *Epic of Gilgamesh* with its flood, mysterious journey, and visions of an afterlife. The tale moves me, in part, because it recounts the story of the Flood in *Genesis*, which is one of humankind's most basic and recurrent memories. For example, the flood is in the *Popol Vuh*, the creation story of the Maya. The *Popol Vuh* flood is not a punishment from God, but rather a remedy for a faulty creation. According to the legend, when humans were first created from mud, the resulting creatures could not see and dissolved if caught in the rain. So God destroyed them and tried again. In his second attempt, God made men out of wood, an experimental creation that turned out somewhat better than the mud men. The wooden men (or "doll people") could walk, talk, and have children, but they had no minds, hearts, or souls. God was once again disappointed with his creations, so he caused a "terrible flood" to cleanse the earth of his mistake, and then the "the creatures of the forest came into the homes of the doll-people." The wooden people scatter into the forest. Their faces were crushed, and they were turned into monkeys, which is why modern monkeys have faces reminiscent of humans.

Plato also gives an account of the flood in the dialogue, *Critias.* According to Plato, the island of Atlantis and other distant Mediterranean civilizations were completely destroyed by earthquakes and floods around 11,400 years ago. Of course, if we took this date literally, it would make the antediluvian civilization a part of the Ice Age. According to Plato, the post-flood survivors lived in the mountains as herdsmen and shepherds.

For a final flood myth, consider the Nez Percé Native Americans of the Palouse, located in what is today Washington state, who also have a flood story in which humans survive by climbing a mountain. The Hopis, Aztecs, and Incas all have flood myths in which God drowns the world and leaves only a few survivors. Could the flood myth be pervasive because many cultures developed in areas where water was plentiful, for example near rivers, seas, or oceans, which are likely to produce a flood at some point in the memory of inhabitants or their ancestors?

Aside from the flood story, *Gilgamesh* fascinates us today—so many centuries after the unknown author's death—because the tale focuses on the mystery of death and immortality. In *Gilgamesh*, Siduri the barmaid tells Gilgamesh, "You will never find that life for which you are looking. When the gods created man, they allotted to him death, but life they retained in their own keeping."[36] Just as in the Homeric epics, humans must die while gods live on in another plane of existence.

I have mentioned that Kübler-Ross found that dying people could accept death most easily if they could look back and feel that they had not wasted their lives. Similarly, Siduri, an enigmatic woman, living in a place "where east and west were confused," advises Gilgamesh to make the most of his mortal life: "As for you, Gilgamesh, fill your belly with good things; day and night, night and day, dance and be merry, feast and rejoice. Let your clothes be fresh, bathe yourself in water, cherish the little child that holds your hand, and make your wife happy in your embrace; for this too is the lot of man."[37]

Gilgamesh can't accept this. Like Elisabeth Kübler-Ross and William James, he yearns to transcend death. As we discussed, in the end of the epic, immortality is given to Gilgamesh in the form of a plant. (Many modern-day shamans have found that DMT-containing plant brews from the Amazon often give people the feeling that they will never die.) Unfortunately for Gilgamesh, a snake eats the magic plant before Gilgamesh can take a bite. Gilgamesh awakens to find the plant gone; he falls to his knees and weeps.

For Gilgamesh, the afterlife contains two separate pocket universes. We already mentioned one that was a depressing place, where feathered dead people wander in darkness and dust. However, the Gods live in a different realm, under the mountain, near the Sea of Death. Alas, this godly place does not seem to be the paradise of the New Testament, because it is mostly empty and the place where the sun goes at night when it descends.

Before we leave Gilgamesh and his kin, I should emphasize that scholars feel that Gilgamesh was an actual historical king who, in 2700 BC, ruled the city of Uruk in Babylonia, which was located on the River Euphrates in modern Iraq. In 2003, archaeologists in Iraq may have found the lost tomb of King Gilgamesh. A German-led expedition discovered what is thought to be the entire city of Uruk including the last resting place of its famous king. In the *Epic of Gilgamesh*, Gilgamesh is described as having been buried under the Euphrates, in a tomb built when the waters of the ancient river parted following his death. The archeologists found a city and a tomblike structure in the middle of a former course of the Euphrates River. Could it be the actual tomb of Gilgamesh? Unfortunately, the archeologists had to stop their work when the war with Iraq started, and further excavations are needed to verify the astounding hypothesis.[38]

In many ways, King Gilgamesh is the ultimate chameleon. At the beginning of the *Epic*, Gilgamesh is portrayed as a selfish tyrant who oppressively rules his grand city. In the end, when he returns from his quest, he is a changed man with new interests, a kindly disposition, and a fresh perspective on life. He leaves a mark on the world and his descendants: "He brought back the ancient, forgotten rights, restoring the temples that the Flood had destroyed, renewing the statues and sacraments for the welfare of the people and the sacred land."[39] Stephen Mitchell comments on Gilgamesh's spiritual awakening:

> When Gilgamesh leaves his city and goes into uncharted territory in search of a way beyond death, he is looking for something that is impossible to find. His quest is like the mind's search for control, order, and meaning in a world where everything is constantly disintegrating. . . . Not until Gilgamesh gives up on transcendence can he realize how beautiful his city is . . . When the mind gives up on its quest for control, order, and meaning, it finds that is come home, to reality, where it has always been.[40]

True journey is return.[41]

Genius is nothing but a power of sustained attention.
          –William James, *Talks to Teachers on Psychology*, 1899

Myth is about the unknown; it is about that for which initially we have no words. Myth therefore looks into the heart of a great silence.

          –Karen Armstrong, *A Short History of Myth*

Gods die with men who have conceived them. But the god-stuff roars eternally, like the sea, with too vast a sound to be heard.

—D. H. (David Herbert) Lawrence, *The Plumed Serpent*

If you've got yourself an early 1950s movie about a killer brain, it either ought to fly around or have grown to gargantuan size, preferably both. The only thing the brain in this one does is sit in a bunch of dirty water in a fish aquarium!

—Jon Wagner, "MonsterHunter Review of *Donovan's Brain*"[42]

Of the religious groups, there are some that have a much harder time [with death] than others. The Jewish people have a terrible issue about death. I tried to find out why. . . . One Rabbi advised the Jews, "You will survive in [your progeny's] memory." Well, after a hundred years, nobody *remembers* you. . . . If you have not concretized your concept [of spiritual faith], then you have a heck of a time with death.

—Elisabeth Kübler-Ross[43]

I've never understood why so many people. . . find [the idea of] oblivion [death] totally dreadful. If you're oblivious, that implies no experience and, of course, no experiencer either. How can you fear or even resent what you will never experience. . . . We all go there every night, between dreams, and it doesn't hurt at all.

—Robert Anton Wilson, in David Jay Brown's *Conversations at the Edge of the Apocalypse*

If your head is cryogenically frozen today, you will be alive in 2100. Your mind is a pattern of activity in your brain. The ability to induce that pattern is encoded primarily in your neurons—neuron type and connectivity. Freezing a brain today in liquid nitrogen destroys many things, but seems to preserve this type/connection information. By 2100 we should be able to scan this information from a frozen brain. If we scan your brain and then build and run a computer simulation of it, someone who remembers being you would wake up and feel alive.

—Robin Hanson, "Fourteen Wild Ideas: Five of Which Are True!"[44]

While 90% of Americans believe in God, only 43% of psychiatrists do. . . . Depressive people tend to be both realistic about life and what is going on around them. I have come to the conclusions that depression is about seeing the world (i.e., *dunya*—the earth-plane if you will) as it is. . . . In order to get out of this depression, one has to get a sense of the other world—the *Akhira* (the afterlife) and also a sense of the *ghaib* (the unseen worlds).

—Psychiatrist Joel Ibrahim Kreps, *Islamica*[45]

FOUR

THE MATRIX, QUANTUM RESURRECTION, AND THE QUEST FOR TRANSCENDENCE

In which we encounter H. P. Lovecraft's "The Whisperer in Darkness," Greg Egan's *Permutation City*, Larry and Andy Wachowski, brains made of bicycle parts, the North American Vexillological Association, Jell-O minds, autistic simulacrums, zombies, actuopalynologists, brain pseudomorphs, cosmic onions, the location of the soul, hyperspecialization, René Descartes, dreaming, the Tajal people, Marilyn vos Savant, *The Matrix*, Frederik van Eeden, lucid dreaming, simulating reality, multiverses, artificial life, gebits, Ray Kurzweil, Emily Dickinson, Digital Philosophy, Stephen Wolfram, Robert Heinlein, *The Truman Show*, George Berkeley, consciousness, Robin Hanson, "peas and carrots," uncommon psychiatric disorders, Nick Bostrom, Process Physics, quantum resurrection, and quantum immortality.

Neo: This isn't real?
Morpheus: What is real? How do you define real? If you're talking about what you can feel, taste, smell, or see, then real is simply electrical signals interpreted by your brain.

—*The Matrix*

Questioning Reality

Whenever I pump gasoline at the local Shell gas station in Shrub Oak, New York, I look around at the nearby people and environment, and I pretend I'm a film director, trying to capture a natural looking scene.

I scan my surroundings. Over there is a woman with long hair getting coffee. Across the street are two truck drivers, talking and waving their arms. The leaves swirl. A child laughs. Birds chirp.

If I were to make a movie, would these people be extras, paid to act naturally, or would they just be random passersby? I look at people on the sidewalk now and see a family with a dog and stroller—followed by a woman in a short paisley skirt and a shirt that says "Ljubljana" in aquamarine letters.

I also think about the kind of music I'd use in my movie to help set the mood. Perhaps songs like "Building a Mystery" by Sarah McLachlan, "Ashes are Burning" by Renaissance, or the fantastic percussion in Peter Gabriel's "The Feeling Begins" that was used in the movie *The Last Temptation of Christ*.

I look to my left. A traffic light turns green. Cars drive by along Route 6 like insects fleeing. In the window of the diner are people eating breakfast. A little girl with a plastic purple pachycephalosaurus skips by. A blackbird cries overhead. I think, *everything looks so real*. And why not? It *is* real.

But what if all this lovely play of lights and birds and people is a facade? Many contemporary movies question our perceptions of reality. What if this is a dream? What if everything I'm watching is just a simulation of reality, like the world in the movie *The Matrix*?

Few of the 1950s science-fiction movies asked us to question reality. For example, *The Brain from Planet Arous*—while touching on the issue of demonic possession—focused more on providing a practical warning for modern society. Humans had rapidly gained technological power but had to be careful to change mental outlooks gradually and naturally. The leading man in the movie was suddenly imbued with powers such as the ability to cause a flying airplane to explode simply by staring at it. Aliens and enemies were all around us. Technology was dangerous. Fear of advanced technology played a central role in movies of the late 1950s.

More recent movie themes make viewers wonder if reality is nothing more than an artificial construct. As we learn more about the universe, and are able to simulate complex worlds on computers, even serious scientists begin to question the nature of reality. Physicists ponder multiple dimensions, parallel universes, and wormholes connecting different regions of space and time. As I discuss at length in *Sex, Drugs, Einstein, and Elves*, powerful drugs like DMT (dimethyltryptamine) cause many of its users to believe that there is more to reality than they had imagined in their unaltered brain states. Reality-questioning movies have multiplied rapidly in recent years—*The Cell, The Matrix, The 13th Floor*, and *Vanilla Sky*, just to name a few. In *The Truman Show*, the main character discovers his world is simulated, and he is living on a complex and massive set for the benefit of a TV audience. In *The Matrix*, our world is an elaborate simulation, a false earth in which our "minds" live—while our bodies are immobilized and kept alive with feeding tubes.

The idea that reality is multilayered—with or without an identifiable ultimate level or reality—is a popular concept in movies that deal with computers, dreams, the afterlife, and altered mental states.[1] Other examples of films with multiple realities include *Neon Genesis* (1997), *Fight Club* (1999), *Being John Malkovich* (1999), *The Sixth Sense* (1999), *Memento* (2000), *Minority Report* (2002), and *Spirited Way* (2002).

I still dream about directing my own movie, starting with an ordinary scene at the local gas station and quickly followed by the main character seeing a flash of movement in his peripheral vision. And then I'd launch into a scene from H. P. Lovecraft to really jolt the reader:

> They were pinkish things about five feet long; with crustaceous bodies bearing vast pairs of dorsal fins or membranous wings and several sets of articulated limbs, and with a sort of convoluted ellipsoid, covered with multitudes of very short antennae. . . . The creatures were a sort of huge, light-red crab with many pairs of legs and with two great bat-like wings. . . . They sometimes walked on all their legs, and sometimes on the hindmost pair only, using the others to convey large

objects of indeterminate nature. . . . A detachment of them waded along a shallow woodland watercourse three abreast in evidently disciplined formation.[2]

Lovecraft was certainly lavish with his writing, wasn't he? He received no serious attention during his lifetime, yet today he has a cult following. Many people around the world consider his tales of cosmic parasites, ventricumbent Gods, and mysterious dreams to be works of genius, and they carefully study his every word. His stories often deal with alternate realities.

Greg Egan's Permutation City

Let's return to the notion of simulated reality. Perhaps the best place to start is with Greg Egan's *Permutation City*, my favorite treatise on the ramifications of artificial worlds and the copying of human minds. Science-fiction author Egan is another chameleon, a novelist and computer programmer with a bachelor of science degree in mathematics. From the age of six, he always imagined having a career as a professional scientist, and later in life he worked as a computer programmer at a medical research institute associated with a hospital in Sydney, Australia. Before his writing career took off, he was an amateur filmmaker. Music has inspired some of his writings, and his story "Beyond the Whistle Test" revolves around scientists who use neural maps to design advertising jingles that you literally can't forget.

Permutation City takes place around the year 2050, when it is possible to make copies of people by using computer software to simulate various aspects of their biology, from brain cells to the retinas. "Less important" organs are not simulated at the cellular level because this accuracy is not necessary to give the copy a realistic sense of being alive. When the simulated brain is run in the computer, the mind has access to all the memories of the real person and can interact with other copies and objects that share the artificial environment.

In other words, Egan is describing electronic cloning of people in an era where computers possess the ability to make perfect replicas of a person's mind, and place this mind into a copy of the person who resides in a virtual world. Due to computer processing limitations, the simulations are usually run at slower rates than similar physical processes run in our real world. If future scientists could simulate you completely, at some suitably low level of physics, would the simulation *be* you in some sense? In *Permutation City*, several rich people seek immortality by uploading their minds to a grand, simulated city that has access to effectively unlimited computer power.

The novel's most interesting character, Peer, has his mind uploaded into a software simulation of the huge and wondrous city. However, because he is not among the wealthy people who live their virtual lives together in social communities, his mind hides in the city's software infrastructure where Peer can only *observe* the city's inhabitants but not interact with them. The good news is that this is a kind of immortality for him because as long as the city simulation runs, he is conscious, and there is no reason that the simulations should ever stop. The bad news is that no one can ever see him or talk to him because he did not have the money to pay for this feature. Luckily, he can simulate his own pockets of reality at the edges of the synthetic city. You can think of Peer's simulated environments as simulations *within* the city simulation. Because he is immortal, he decides to engineer his mind so that he can be totally happy and at peace by doing repetitive but satisfying tasks. For example, he constructs a plantation of sugar pine trees using information from a genetic library and biochemical models. The years pass as he remains confined to a pleasant workshop, making table legs on a lathe, one after the other. He never tires. He is alone, but not lonely. He is never bored because this emotional component is removed from his mind. You and I, with our current mind states, would find the task monotonous.

In the novel, we come to learn that Peer's workshop abuts a warehouse full of table legs that he has made—162,329 legs to be exact. Peer finds it pleasurable to imagine that he may someday reach the

200,000 mark, although he knows this is unlikely because he had instructed the computer running his simulation to impose new vocations at random intervals. For example, before he had become a woodworker, he had passionately read all of the higher mathematics texts in the computer's main library and then personally contributed new results on group theory. Alas, the city inhabitants would never see his cutting-edge mathematics because he was a stowaway in the software. However, the fact that his work would never be seen by anyone did not make him sad because he had engineered his own mind to be perpetually at peace, even in his isolation. Before he had embarked on his math studies in his little corner of the simulated world, Peer explored literature, biology, and art:

> He'd written over three hundred comic operas, with librettos in Italian, French and English—and staged most of them, with puppet performers and audience. Before that, he'd patiently studied the structure and biochemistry of the human brain for sixty-seven years; towards the end he had fully grasped, to his own satisfaction, the nature of the process of consciousness. Every one of these pursuits had been utterly engrossing, and satisfying, at the same time.[3]

If *you* had hundreds of years to spend alone in a simulated world with essentially limitless resources, would you begin your adventure by trying to design artificial companions? Would you re-engineer your mind to be happy in such isolation? Perhaps you'd never be bored if you could spend your days in a virtual Library of Congress, which today has over 128 million items and over 520 miles of shelves! Would you become an expert on the Punic and Peloponnesian Wars or fluent with gagaku, the East Asian music popular during the fifth to ninth centuries? Perhaps you would want to be the universe's most knowledgeable:

- Vexillologist—one who studies flag history and symbolism (yes, there is a North American Vexillological Association for flag devotees)
- Sigillographist—one who studies wax seals
- Selenographist—one who studies physical features of the moon
- Actuopalynologist—one who, according to Professor Owen Davis of the University of Arizona, is a person who "studies extant palynomorphs [spores or the remains of microscopic creatures], which are either living, still retain their cell contents, or whose cell contents have been removed by maceration. Actuopalynology includes mellisopalynology, pollination ecology, aeroallergy, and criminology."[4]

What obsessions would develop in an eternal simulated world with few or no companions with whom to interact? Maybe a few weirdoes obsessed with biology would spend their time performing virtual dissections on the Tyrolean Ice Man, the five-thousand-year-old man found high in the Otztal Alps on the Austrian-Italian border. My favorite writing on the subject is Dr. Tim Holden's blockbuster chapter, "The Food Remains from the Colon of the Tyrolean Ice Man" (in Keith Dobney and Terry O'Connor's *Bones and the Man: Studies in Honor of Don Brothwell*, Oxford: Oxbow Books, 35–40, 2002). In fact, if your afterlife existed in a simulated environment, with just a few software tweaks you could read and enjoy the entire *Bones and the Man* book, which includes the following intriguing chapter titles:

- "Dental Anthropology 30 Years On"
- "Myopia and Nutritionally-inhibited Cranio-facial Growth"
- "The Food Remains from the Colon of the Tyrolean Ice Man"
- "Brain Pseudomorphs: Grey Matter, Grey Sediments, and Grey Literature"
- "Was Bucephalus' Burial for Real? Recent Finds of Horse

Burials in King Philip's Tomb at the Great Tumulus of Aigai,
Greece"
• "The Rat Race: On the Quest for the Oldest Commensal
Rodent"

Given an eternity, traditional hobbies may morph from general to
specialized, for example, from a general curiosity in anthropology to
interest in the bacterial ecology within the large intestine of the Ice
Man. Perhaps more detailed simulations could be run to explore
deeper aspects of microbiology and ecology.

I include these Ice Man digressions for three reasons: 1) to give you
an idea about how people might choose to fill their days if they were
immortal software simulations; 2) to give an indication of the degree
of hyperspecialization taking place in the realm of knowledge today,
which reinforces our Chapter 2 discussion on specialization; and 3) to
see if the stodgy publisher will allow me to include such zany digressions for which no reasonable justification exists. Most publishers typically excise ramblings or "information dumps" that slow the flow of
the book, even if I personally find the digression humorous and totally
fascinating. (Incidentally, the "brain pseudomorph" mentioned in the
previous list refers to an object that forms when an animal dies in mud
and the soft tissues of the brain are slowly replaced by minerals.)

Let's return our attention to living in simulated worlds. If we are
someday able to replicate our minds by uploading them as software
models running on computers, a great many laws must be passed to
ensure a civil society in the virtual world and to minimize suffering.
Robin Hanson, an assistant professor of economics at George Mason
University, notes:

As with software now, illicit copying of uploads might be a big
problem. An upload who loses even one copy to pirates might
end up with millions of illicit copies tortured into working as
slaves in various hidden corners. To prevent such a fate, uploads

may be somewhat paranoid about security. They may prefer the added security of physical bodies, with "skulls" rigged to self-destruct on penetration or command. And without strong cryptography, they may be wary of traveling by just sending bits.[5]

Are You Dreaming?

You may not believe that humanity will ever be able to simulate realities so accurately that the artificial worlds will be "actual" realities on par with the world in which we live. You may not believe that we'll be able to plug our brains into machines to help us dream entirely new realities or that we'll download ourselves to computers where the world looks perfectly real but is no more real than a dream. Nevertheless, we can still ponder a related question: Could our world be just a dream? Could you be dreaming right now? This kind of question has been pondered since the time of philosopher René Descartes (1596–1650) and some of the ancient Greek philosophers. The Tajal people of the Philippines believe that the soul leaves the body during sleep and enters a dream reality as real as our own, which is why the Tajal punish people for awakening a sleeper–premature awakening can destroy the soul.

In various folk tales, when a person falls asleep, a small animal such as an insect leaves the person's mouth and returns when the person is ready to awaken. The stories imply that the insectile form carries the soul and that the person may die if the creature cannot gain reentry to the sleeping human.

One night, after reading about insectile soul-carriers, I had my own dream of advanced mantislike aliens that descended to Earth. They abducted people and asked, "What is the best way to determine if you are dreaming?" If your reply is sufficiently insightful, they will set you free. What is your response? In short, how can you tell if you are dreaming?

This question reminds me of the movie *The Matrix*. In one particularly memorable scene, Morpheus says to Neo, "Have you ever had a dream, Neo, that you were so sure was real? What if you were unable to wake from that dream, Neo? How would you know the difference between the dream world and the real world?" Various intellectuals have pondered this dream question as it relates to our everyday life. Marilyn vos Savant was once listed in the *Guinness Book of World Records* as having the highest IQ in the world—an awe-inspiring 228. A reader asked her, "How can you tell if you are dreaming?" Her response was, "If you're wondering if you're dreaming, you're dreaming."

William Poundstone in *Labyrinths of Reason* has a recipe for making sure you are not dreaming. He suggests you keep a book of limericks that you've never read at your bedside. Whenever you want to determine if you are dreaming, open the book at random and read a limerick. He reasons that in real life, and certainly in dreams, it is impossible to quickly create a perfect limerick.

Clearly, we have difficulty determining when we are dreaming when we are in fact dreaming. Thus, if we lived in a simulated reality like in the movie *The Matrix*, the simulators could easily hide this fact from of us if their simulations were sufficiently advanced. After all, our dreams often contain bizarre anomalies, and yet we don't realize we're dreaming. It's as if the brain suppresses this realization for a purpose. But for what purpose? Perhaps during sleep, those parts of the brain responsible for logical thinking are damped, and as a result, we interpret dreams more symbolically than in terms of cause and effect. Sam Harris remarks in *The End of Faith:*

> Our waking and dreaming brains are engaged in substantially the same activity; it is just that while dreaming, our brains are far less constrained by sensory information or by the fact-checkers who appear to live somewhere in our frontal lobes. This is not to say that sensory experience offers us no indication of reality at large; it is merely that, as a matter of experience, nothing arises in consciousness that has not first been structured, edited, or amplified by the nervous system. While this gives rise

to a few philosophical problems concerning the foundations of our knowledge, it also offers us a remarkable opportunity to deliberately transform the character of our experience.[6]

Dutch novelist, poet, physician, and dream researcher Frederik van Eeden (1860–1932) once performed a marvelous experiment in which he tried to leave a mark on the "real world" while still in a dream. In particular, he attempted to make a mark on his forehead in his dream so that the mark would persist into the real world. At first, he thought he succeeded, but then something strange happened: "In February 1899, I had a lucid dream, in which I made the following experiment. I drew with my finger, moistened by saliva, a wet cross on the palm of my left hand, with the intention of seeing whether it would still be there after waking up. Then I dreamt that I woke up and felt the wet cross on my left hand by applying the palm to my cheek. And then a long time afterwards I woke up really and knew at once that the hand of my physical body had been lying in a closed position undisturbed on my chest all the while."[7]

In other words, van Eeden had experienced a "false awakening," moving him from one level of awareness—he knew he was dreaming—to another level—thinking he was awake when he was actually still dreaming. Sometimes, dreamers traverse several realities, or false awakenings, before becoming truly awake, causing philosophers to wonder if everyday consciousness is not still some sort of dream within a dream.

Richard Linklater's 2001 animated film *Waking Life* is a fictional example of a deeply nested set of dreams, and an exhilarating meditation on the nature of reality. As you watch *Waking Life*, you'll begin to question your own reality. One scene focuses on two men discussing the nature of film as an art form. Gradually, the viewer's perspective widens, and we see that the men are characters on a movie screen in a theater with the main character sitting in the theater watching. And he is dreaming he is watching the movie. And we are in our homes, watching him dreaming the watching. These layers remind me of Robert Pirsig's beautiful quote in his novel *Lila*: "He watched her for a long time and she knew that he was watching her and he knew that she knew he was watching her, and he knew that she knew that he knew; in a kind of regression of images that you get when two mirrors face each other and the images go on and on and on in some kind of infinity."

Waking Life makes a number of interesting observations, including the fact that a sleeping dreamer cannot turn off the lights in dreams. For example, when I have lucid dreams in which I am aware that I'm dreaming and can often exert some measure of control of the dream, I have never been able to turn off the lights with a light switch. I've polled numerous colleagues, and none of them has recalled being able to accomplish this.

Have you ever wondered about what kinds of acts you cannot perform in dreams? For example, I find it very difficult to make phone calls, send computer instant messages, or read a book while in dreams. I can't seem to focus on the sentences and digest their meanings. One common test to determine if you are dreaming involves looking at a text in the dream, gazing away, and then looking at the text again. Usually, the text is not the same during the second gaze. A few friends have also found it impossible to read a full sentence in dreams. Others observe that they are unable to penetrate their bodies with their hands, even when they are lucid and have control over other aspects of the dream. Dream researcher Chuck Gaydos writes to me:

> I can sometimes read individual words or short phrases in dreams but nothing lengthy. Even if I remember reading a word, I can never remember it after I wake up, except once when I remembered dream reading and wrote down "The Iron Art" right after waking up. I have no idea what it meant.
>
> In one lucid dream, while levitating, I tried to reach my hand into my chest, figuring I should be able to reach through

my body since it was only a dream. My fingertips stopped at my shirt.[8]

Chuck and I have noticed that reflections in dream mirrors often present anomalous reflections. For example, Chuck has raised his hand and observed the reflection might not move his hand quite as high. Sometimes, a person's reflection disappears completely. Another friend, April, has never been able to spend money or even have money in her possession while in a dream! She has had dreams where she's won a jackpot in Las Vegas, but never received the money—or she is in a gift store and realizes that she has no money. She writes, "I've never even seen a coin, dollar, or any currency in my dreams!"

On the other hand, several colleagues I polled sent me lists of things they *can* in fact do in dreams. These include:

- The ability to experience taste in dreams. (One person notes that if he eats something in a dream it is almost always something like chocolate cake or ice cream. A woman also tells me she eats only sweets in dreams.)
- The ability to produce "high-fidelity, elaborately produced music of much higher quality than possible to create in actual life, even in largely unfamiliar musical genres."
- The ability to read in dreams, "but I can never really bring any text I encounter into the waking world, and the text changes when I get distracted by something else."
- The ability to type on a computer keyboard and view the results on screen. "However, I don't offhand remember switching the computer on, and I also don't remember the computer ever crashing."

My high-IQ colleague James Platt spends some of his lucid dreams looking up at the sky. He finds the relative lack of detailed visual information in the sky causes his mind to "form startling visual patterns in the blue or black spaces, and almost always causes some sort of strange event, not always to my liking, but always interesting."

Platt has also had dreams within dreams. His nested dreams involve little content or motion, and they last for less than a minute. James describes the nested dream as a "frozen idea or memory, as if this dream self, being just a fraction or partial interpolation of the waking self is not capable of producing another elaborate dream, because the dream self has access only to a portion of the brain or mind."

When I am lucid, I often don't realize that the person I am with in the dream is just a manifestation of my mind. For example, I have had dreams in which I enter a club playing ancient Persian music. I am with an Iranian friend and know that I'm dreaming. While listening, I'm surprised that my mind can manufacture the music with sufficient realism that my friend feels that the music is authentic and recognizes the music. In other words, I know I am dreaming but do not realize that the person I am with is simply a projection of my mind.

Consider a related example. My cell phone once rang in a dream, and I realized that the ring tone was not the usual one produced by my phone. I turned to my friend in the dream and explained to him that this phenomenon probably occurred because one ring tone was used in the dream world and another tone produced in the real world—but, during the dream, I didn't seem to realize that my friend was a product of my imagination.

In order to encourage the emergence of lucid dreams, Stephen LaBerge, founder of the Lucidity Institute, uses electronic devices like the NovaDreamer, a sleep mask that emits a red light when the user starts to dream. The dreamer sees the light through his eyelids, realizes he is dreaming, and thereby increases his chances of becoming lucid.[9] People who use this device can gain remarkable control of their dreams, exploring new realms of thought and imagination.

LaBerge also uses a host of simple methods to stimulate the occurrence of lucid dreams, such as rubbing hands together in dreams or spinning the body, and taking capsules of the supplement galanta-mine.[10] The NovaDreamer mask also has a Reality Check button that activates a blinking light. If the user is not sure if he is awake or asleep, all he needs to do is press the button. If the light blinks, the user is probably awake. If the light does not blink, the user is probably only dreaming that he pressed the button.

We Are Simulations

In *The Matrix*, Morpheus tells Neo that the world in which he lives is a mere simulation running within the collective brains of immobilized bodies. People's bodies are like ancient ants trapped in amber, but their minds are free to fly in an infinite cirrostratus sky. Morpheus says, "You've felt it your entire life, that there's something wrong with the world. You don't know what it is, but it's there, like a splinter in your mind, driving you mad." Ultimately, Morpheus proves to Neo that there is a reality beyond reality that controls all of their lives. . . . It all sounds rather far-fetched, doesn't it? It's a bit hard to believe that your life is a fantasy induced by computer software or by equations running on some machine. However, a number of scientists say this scenario is possible, even likely.

In our own small pocket of the universe, we've already developed computers with the ability to simulate lifelike behaviors using software and mathematical rules. One day we will create thinking beings that live in rich simulated spaces—in ecosystems as complex and vibrant as an Amazonian rain forest. We'll be able to simulate reality itself, and perhaps more advanced beings are already doing this elsewhere in the universe. Huge supercomputers would have the capacity to sim-ulate not just a tiny fragment of reality, but a substantial fraction of an entire universe.[11]

What if the number of these simulations is larger than the number of universes? Could we be living in such a simulation? Astronomer and philosopher Martin Rees suggests that if the simulations outnumber the universes, "as they would if one universe contained many computers making many simulations," then it is likely that *we* are artificial life. He notes that this theory allows for "virtual time travel," because the advanced beings who create the simulation can rerun the past. Rees says in his essay "In the Matrix" (aka "Living in A Multiverse"),

> Once you accept the idea of the multiverse, and that some universes will have immense potentiality for complexity, it's a logical consequence that in some of those universes there will be the potential to simulate parts of themselves, and you may get a sort of infinite regress, so we don't know where reality stops and where the minds and ideas take over, and we don't know what our place is in this grand ensemble of universes and simulated universes.[12]

Astronomer Paul Davies in "A Brief History of the Multiverse" has remarked similarly, "Eventually, entire virtual worlds will be created inside computers, their conscious inhabitants unaware that they are the simulated products of somebody else's technology. For every original world, there will be a stupendous number of available virtual worlds—some of which would even include machines simulating virtual worlds of their own, and so on ad infinitum."[13]

As I mention in *The Möbius Strip*, computer scientists and biologists have already made tremendous progress in creating "artificial life"—lifelike entities with complex behaviors that are simulated using simple rules implemented in computer software. Currently, the study of artificial life reminds me of bacteriology or apiology (the study of bees). Researchers have simulated simple life forms with short life spans, living in simple societies. For example, Professor Tom Ray of the Department of Zoology at the University of Oklahoma created Tierra, a system in which self-replicating machine-code programs evolved by natural selection. Although these creatures were very small, only a few instructions long, they exhibited many behavioral patterns found in nature. Diverse ecological communities emerge whenever many of these and other simulated biomorphic entities interact. These kinds of digital communities have been used to experimentally examine evolutionary competition, punctuated equilibrium, the emergence of complex behaviors, and the role of chance in evolution.

A glance at the *Proceedings of the Ninth International Conference on the Simulation and Synthesis of Living Systems (Artificial Life IX,* MIT Press) reveals the kinds of topics that computer scientists and robotics experts are considering—with such delicious paper titles as:

- "Designed and Evolved Blueprints for Physical Self-Replication Machines"
- "Crawling Out of the Simulation: Evolving Real Robot Morphologies Using Cheap, Reusable Modules"
- "Evolving Flying Creatures with Path Following Behaviors"
- "Evolving Simulated Mutually Perceptive Creatures for Combat"
- "A Comparison of Population Learning and Cultural Learning in Artificial Life Societies"
- "Whatever Emerges Should Be Intrinsically Useful"
- "Complex Genetic Evolution of Self-Replication Loops"
- "Lipidia: An Artificial Chemistry of Self-Replicating Assemblies of Lipidlike Molecules"
- "Ant Foraging Revisited"

In addition to lifelike simulations being run in the brains of computers, scientists also study robotic herds and flocks. For example, James McLurkin from MIT creates autonomous robots that form herds, and in the future these robots may work together and take creative action. British contributors to the University of Essex's "Ultra-Swarm" project are creating a fleet of tiny, robotic helicopters capable of swarming like bees and evaluating their surroundings with a single hive mind. The swarm entities communicate with wireless transmissions, allowing the swarm to move as one entity. According to the researchers, a typical flock of starlings—about 2,000 birds—contains as much brain tissue as a single human.

Dave Payton from the HRL Research Labs in Malibu, California, avoids the need for centralized control of his robots by having his pherobots use virtual pheromones to communicate. When a pherobot finds something of interest in its environment, it transmits a series of infrared pulses that are picked up by its nearest neighbor. Eventually, trails of pherobots flock to the first pherobot to see what it has found.

Craig Reynolds's Boids are computer simulations that flock and school like birds and fish. Craig, who now works for Sony Computer Entertainment, used only three simple rules to govern the Boids's life: 1) steer to avoid getting too close to neighbors; 2) steer to keep on the average heading of the flock; and 3) steer to stay near the average position of the neighbors. Rules about goal seeking and obstacle avoidance can be added, so that the artificial creatures can navigate through a world filled with objects. The resulting behaviors based on these simple rules are remarkable, and it is not hard to imagine that simple rules and simulations known as cellular automata could develop complex societies of reproducing entities, especially when millions or billions of the creatures interact in huge worlds.

Examples of other artificial life forms and ecologies include: Larry Yeager's Polyworld, Kevin Coble's Neoterics, and Karl Sims's multilimbed creatures (MCs) that compete with one another. The MC brains are neural nets with several sensors. Through competition, the creatures evolve intelligent behaviors that would be hard for humans to actually design.

The Avida system—a joint project of the Digital Life Laboratory at the California Institute of Technology and the Microbial Evolution Laboratory at Michigan State University—provides a software platform in which digital organisms breed thousands of times faster than common bacteria and shed light on some of the biggest unanswered questions of evolution. Darwin-at-Home is a planetwide effort to create networked digital ecosystems and to recreate the evolution of life on Earth by permitting computer creatures to evolve (www.darw inathome.org). The Darwin teams hope to observe lifelike evolutionary processes in a virtual or robotic space. Their interactive computational platform is distributed across a large pool of networked computers, which allows people to shape each digital biotic ecosystem.

The field of artificial life exploded in the twentieth century after British-born mathematician John Conway invented a simple "game" in which primitive, simulated life forms, based solely on mathematical rules, evolved before scientists' astonished eyes. Conway was interested in mathematics from a very early age and could recite the powers of two when he was only three or four years old. As an older boy, the fields of mathematics, astronomy, arachnology, and paleontology all occupied his continually roving mind.

In 1970, Conway devised a game called Life, which is the most famous of cellular-automata life forms. In this simple simulation, cells live or die on a two-dimensional grid of cells when they follow just two rules: 1) A cell is turned on (is born) if three of its neighbors are turned on, and 2) A cell remains on (lives) if two or three of its neighbors are also on; otherwise it is turned off (dies). These simple rules control the birth, survival, and death of any cells through time. Sometimes, entities or shapes composed of a collection of cells evolve and move around the checkerboard universe while maintaining their overall shape, just like a paramecium moving though a drop of water on a microscope slide. In fact, some forms evolve that are able to maintain their shape and spawn other shapes that are then able to explore the environment, essentially simulating the act of reproduction. If lifelike phenomena emerge using such simple rules, we can expect that certain rules executed on checkerboard universes could spawn complex societies, given sufficient time and a sufficiently large world on which to evolve. Similar kinds of checkerboard creatures have the potential to reproduce, contract diseases, exchange information, learn to use tools, build shelters, and wage wars. Hopefully, they will also learn to dream.

Greg Egan in *Permutation City* suggests that medical imaging technology will continue to improve, and that by 2020 it will reach the point where individual neurons can be mapped and the properties of individual synapses measured, noninvasively. "With a combination of scanners, every physiologically relevant detail of the brain could be read from the living organ—and duplicated on a sufficiently powerful computer."

Egan suggests that, at first, only isolated neural pathways will be modeled—"portions of the visual cortex of interest to designers of machine vision, or sections of the limbic system whose role had been in dispute." He writes from a vantage point in the future:

> These fragmentary neural models yielded valuable results, but a functionally complete representation of the whole organ. . . would have allowed the most delicate feats of neurosurgery and psychopharmacology to be tested in advance. . . . In 2024. . . a Boston neurosurgeon ran a fully conscious Copy of himself. . . . The first Copy's first words were: "This is like being buried alive. I've changed my mind. Get me out of here."[14]

According to Robert J. Sawyer, author of the novel *Mindscan*, in the future we will become immortal by first scanning our brains—"every neuron, every dendrite, every synaptic cleft, every interconnection. . . , [every] neurotransmitter level at each synapse." The brain-scan information is loaded into durable artificial brains that "congeal out of quantum fog, forming a nanogel that precisely duplicates the structure of the biological original." Sawyer also suggests that the enormous amount of data involved in brain scans may be reduced because the brain information is fractal, often with the same kinds of patterns repeated at different levels of resolution. Thus, the data is easy to compress.

To ensure the mental health of the people undergoing mind

transfer, *Mindscan* scientists find that these artificial brains must be pre-installed in robotic bodies before the person wakes up. Initial experiments that had used isolated brains—which were unable to communicate, move, or have sensory input—showed neural patterns indicative of terror and insanity that arose from an unnatural separation from the real world.

I can imagine a scenario in which I would go to sleep at night, be scanned and uploaded to the nanogel brain while asleep, and then my original body and mind are destroyed before awakening in the morning. For continuity, the immortal me could even be placed in the same bed so that when I awakened in the morning and stretched my new arms, I would have little emotional shock. Just imagine how it would feel to go to sleep and step out of bed as an immortal the next day. However, Sawyer does not follow this scenario but rather has the original "you" living in unsurpassed luxury in a retirement village on the far side of the moon. The moon's reduced gravity decreases the incidence of broken bones from falls, and gives the residents strength even with their weakening muscles as they naturally age. Eventually the moon "you" dies, while the gel-brain in its hot android body lives forever.

Futurist Ray Kurzweil says that the available scanning resolution of human brain tissue—both spatial and temporal—is doubling every year, and so is our knowledge of how the brain works.[15] He also says that the brain is not one entity or one big neural net, but rather it is composed of several hundred different regions that we should be capable of understanding and modeling with mathematics. This has already been done for a few dozen regions out of the several hundred, and in twenty years we will have realistic simulations and understanding of all the regions. Michael Gazzaniga, Director for the Center for Cognitive Neuroscience, Dartmouth College, notes, "Both activation studies and very recent studies on connectivity allow for the identification of specific networks that are active in mental states ranging from basic perception to empathy."

In 2005, a team of Swiss neuroscientists began to build a computer model of the neocortical column of a rat, a single circuit of about ten thousand cells, each of which has thousands of connections. According to Susan Kruglinksi's "Big Blue to Build A Brain," rat brain information is downloaded into IBM's Blue Gene supercomputer, which can perform twenty-two trillion operations per second. The team hopes to replicate the entire rat brain and eventually begin modeling human brains. In my novel, *The Heaven Virus*, I mention other recent examples of brain replication. For example, researchers at the Redwood Center for Theoretical Neuroscience in California are also attempting to create a model of the brain that can be tested in the lab and eventually mimicked by a computer. Olaf Sporns, a neuroscientist at Indiana University, has proposed a project on par with the ambitious Human Genome Project to map the neural connections throughout the human brain. The blueprints of these connections, called connectomes, will involve the worldwide effort of brain imaging scientists, anatomists, and computer scientists.

Where Is the Soul?

The notion that we can one day simulate a mind or upload it to a computer assumes a *materialist* view in which the mind arises from brain activity. On the other hand, some people still adhere to the belief of René Descartes, who, four hundred years ago, thought that the mind, or soul, exists separately from the brain, but is connected to the brain via an organ like the pineal gland, which acts as a portal between brain and mind. (The pineal gland is a pea-sized brain organ and is the only symmetrical organ in the brain without a left and right counterpart.) The third-century-BC physician Herophilus dissected skulls and decided that the soul was located in the fourth chamber of the brain. Leonardo da Vinci suggested that the soul sat atop the spinal column. In 1901, surgeon Duncan Macdougall placed a dying tuberculosis patient on a scale. Macdougall reasoned that at the moment of death, the scale should indicate a drop in weight as the soul disembarked. As a result of this macabre experiment, Macdougall measured the soul to be 21 grams. Alas, Macdougall and other researchers were never able to duplicate this finding. These various views of soul and matter represent a philosophy of *dualism*.

To my mind, more evidence exists for the materialist view because virtually every aspect of our thinking, memory, and personality can be altered by damage to regions of the brain, and brain imaging studies can be used to map both feelings and thoughts. For example, injury to the brain's right frontal lobe can lead to a sudden, passionate interest in high cuisine, gourmet foods, and fine restaurants—a condition called gourmand syndrome. Of course, the dualist Descartes might have argued that damage to the brain changes the behavior of a person because it is through the brain that the mind operates. A dualist would note that if we excise the steering wheel from a car, the car will behave in a different fashion, but this does not mean that there is no driver. However, I personally don't find dualism to be a likely explanation for how the brain and mind work. For example, the concept of a disembodied "soul" coming through the pineal gland or Fissure of Rolando seems to be a paranormal leap that we couldn't accept into the realm of science without more evidence.

Scientific studies of memory also support a materialistic view. Long-term memories are stored by chemical and structural changes in neural patterns—including alterations to neuron branching patterns and the number and efficacy of synapses. Researchers at the Department of Neurobiology and Anatomy at the University of Texas have precisely documented these changes in mollusks in their paper "Neural and Molecular Bases of Nonassociative and Associative Learning in Aplysia," and there's no reason to assume that mammalian brains use a substantially different approach to memory storage.[16]

Game companies are currently on the path to synthesizing artificial realities using our own neurons. In 2004, Sony's US patent 267,118 describes a method that provides a first step to a real-life Matrix. The Sony device fires pulses of ultrasound at a user's head to modify neural activity that in turn creates sensory experiences involving sight, taste, and sound. This approach may one day even let the blind see and the deaf hear.

Ray Kurzweil predicts that, within fifty years, humanity will have access to a virtually unlimited amount of computing power through three-dimensional assemblies of nanotube circuitry made from carbon strands. He suggests that a "one-inch cube of nantotube circuitry would be a million times more powerful than the computing capacity of the human brain."[17] According to Kurzweil, by 2050, $1000 of computing will equal one billion human brains. Moreover, we'll be able to create virtual worlds by sending billions of nanobots to reside at positions by every nerve fiber supplying our sensory apparatuses. "If you want to be in real reality, they sit there and do nothing. If you want to be in virtual reality, they suppress the signals coming from our real senses and replace them with the signals that you would have been receiving if you were in the virtual environment."[18]

As we discussed, many science-fiction novels like *Permutation City* and *Diaspora* discuss the possibility of uploading copies of your mind to a computer simulation. You become software running on a machine. Sometimes, I wonder what it would be like to download other people's memories to my mind or to my simulation's mind. With a sufficient number of memory downloads, it becomes difficult to say who *you* really are. What does identity mean if you are 33 percent someone else in terms of memories? Could you keep from going mad? Or as Emily Dickinson wrote,

> And Something's odd—within—
> That person that I was—
> And this One—do not feel the same—
> Could it be Madness—this?[19]

Aside from making us ask questions about the nature of reality, the movie *The Matrix* mostly focuses on the adventures of rebels who discover that their world is merely a simulation created by intelligent machines. Human minds participate in a realistic simulation of the world, a "computer-generated dream world" that is the simulation of the year 1999. Most viewers of the *The Matrix* would probably opt to join the rebels who fight the machines and yearn for a destruction of the Matrix. However, upon further thought, would you really want to do this? Without a rebellion, most of the humans would continue to live happy lives. Is knowing "the truth" worth it, even if the truth does not make one happier? Neo hopes that a world without the Matrix is "a world without rules and controls, without borders or boundaries, a world where anything is possible." But this seems utterly false. Wouldn't the world still be one filled with hunger, fear, pain, and death? What if the greatest numbers of happy minds are produced by living in the Matrix? If I could opt for hundreds of years of subjective time having wonderful adventures, ideas, and relationships in a Matrix—achieving bliss—I'd take it.

The Mysterious Wachowski Brothers

The chameleonic creators of *The Matrix*, Larry and Andy Wachowski, remain shrouded in mystery. They rarely give interviews and, according to *WIRED* magazine, as part of an unusual deal with Warner Brothers, they never have to talk to the press. Born in Chicago to a businessman father and a mother who was a nurse and painter, both dropped out of college after two years and set up a construction, painting, and carpentry business.[20] (Recall that John Cage dropped out of college after two years and Truman Capote never went to college.)

We will speak of Roger Corman and his 1950s movie *Attack of the Crab Monsters* in the next chapter, but I introduce him here because the Wachowski brothers love Corman and his movies. Corman's book *How I Made a Hundred Movies in Hollywood and Never Lost a Dime* was the inspiration for the Wachowski Brothers to get into the movie business. They began to write scripts, and in a few years Warner Brothers gave the green light for *The Matrix*.[21]

Of course, the Wachowski Brothers were not the first to question reality and wonder about the pros and cons of living in a synthetic Matrix. In the 1970s, philosopher Robert Nozick asked in his book *Anarchy, State, and Utopia* if people might actually *want* to be hooked up to an "Experience Machine." The device would give you the experiences of being very happy—with friends, fame, wealth, good looks, success, and anything else that gives you pleasure and bliss.[22] Once attached to the machine, you would forget about your real-world life, and you wouldn't be unhooked from the Experience Machine.

Stop and think. If you had no loved ones in this world to whom you feel attached by love or obligation, would you choose to be wired to the machine? This scenario is similar to being a brain floating in a vat, attached to electrodes. The additional feature of not knowing you are a disembodied brain is probably important to help maintain your sense of peace.

The idea that we likely live in a Matrix-like simulation, or more generally, that the universe can be thought of as a giant computer, is seriously considered by physicists who ponder the Computational Hypothesis of reality. Philosopher David Chalmers explains:

> The Computational Hypothesis says that physics as we know it is not the fundamental level of reality. Just as chemical processes underlie biological processes, and microphysical processes underlie chemical processes, something underlies microphysical processes. Underneath the level of quarks and electrons and photons is a further level: the level of bits. These bits are governed by a computational algorithm, which at a higher-level produces the processes that we think of as fundamental particles, forces, and so on.[23]

Ed Fredkin, former head of MIT's Laboratory for Computer Science, has also likened the universe to a giant computer. His Digital Philosophy might be considered an atomic theory carried to a logical extreme where all quantities in nature are finite and discrete. This means that quantities can be represented by integers and that our lives can be modeled by computations. Fredkin suggests that our reality may be a simulation on some alien's "computer" in a parallel universe. Based on Fredkin's works, science-writer Keay Davidson describes us as "passive observers of a nonstop TV show pumped into our minds, like a 'virtual reality' simulation from which we can't escape."[24] Fredkin actually believes that computer bits are more concrete than photons or electrons. Thus, this proposed digital universe is "not a 'simulation' of reality; it's not something that 'pretends' to be reality. It *is* reality."[25]

More recently, Stephen Wolfram in *A New Kind of Science* suggests that reality is based on cellular automata, with bits turning on and off according to simple rules. If the physical world really is a computation obeying certain rules that might be expressed in lines of a computer program, perhaps the bit world can supply us with the same kind of spiritual sustenance and hope for immortality as did Utnapishtim in the *Epic of Gilgamesh*. As we have discussed, in order to become immortal, Gilgamesh journeyed far to visit Utnapishtim and his wife, the only humans to have survived the Great Flood and to whom the gods granted immortality. In a world of bits, immortality may become possible if the lines of computer code can be changed or replayed. Perhaps the computer program for reality is run in the mind of God. If God plays with the bits, there is always the hope that He can interface with us by running replicas of Himself in the simulated reality, or by scattering bits of himself at strategic locations throughout our bodies and universe, like stars sparkling in the galactic fabric.

Brains Made of Bicycle Parts

Robert Heinlein's story "They," originally published in 1941, describes a patient in a mental hospital who believes that the world was created to deceive him. In fact, he believes that the entire world is a sham, constructed around him to prevent him from recalling his true identity. He suspects that most people he meets are automatons of some kind, programmed to distract him, but he feels that some people are real and responsible for the deceptive environment. He seems clearly insane.

However, the reader gradually comes to believe that the man may be correct, especially when he goes to the back of his house to get an umbrella and discovers that it is sunny at the back of the house while it is raining in the front. We soon learn that the world is indeed an illusion created by godlike beings to keep him ignorant of his own true powers. More recently, the movie *The Truman Show* follows the life of Truman Burbank who lives in an artificial world filled with actors meant to keep him from learning that his world is a deception.

Sometimes I wonder what our ancient philosophers would have thought of the movie *The Matrix* or ideas that our world is simply made from digital computations. Irish philosopher George Berkeley (1685–1753) defended the view that the physical objects that we sense are nothing more than collections of sensations. There is no "real" object or matter that is the foundation of the perceived object and the cause of our perception. According to Berkeley, your perception of a chair is an idea that God's mind has produced in yours, and the chair continues to exist in your room when you are elsewhere simply because God is always there. Berkeley also believed that knowledge of the world may be made clearer by quieting all words and thoughts so that they do not interfere with our perceptions. Perhaps Berkeley would have thought that the Matrix gives rise to perceptions in the same way God does in our own world.

If our thoughts and consciousness do not depend on the actual substances in our brains but rather on the structures, patterns, and relationships between parts, then brains embodied with other materials or in software could think. Today, British physicist Martin Howard has already created functioning computer logic circuits in LEGO (www.goldfish.org.uk), and Denis Cousineau of the Université de

Montréal has built more sophisticated computers with memory features, also from LEGO parts. Could these be the first steps for imagining a LEGO mind? If we believe that consciousness is the result of patterns of neurons in the brain, our thoughts, emotions, and memories could even be replicated in moving assemblies of bicycle parts. The bicycle minds would have to be very big to represent the complexity of our minds, but nevertheless could be created, at least in theory. However, if you were to walk around inside a sun-sized bicycle mind, you could never point to the conscious entity. Run your fingers lingeringly over the gears or wheel spokes, and the soul is not there. Kick the nearest pedals and pluck the oil-stained chains, but you cannot feel the mind. The mind emerges from the collection of interacting parts and patterns. And I have no doubt that the bicycle mind could love and dream. Even bicornuate bicycle brains can become hyperchameleonic, translinguistic, and religious—religious enough to worship God and pray.

One way to see how a conscious brain may be created from electronic circuits or bicycle parts is to create the brain gradually. For example, imagine that your brain has a small defective portion that we can replace by electrical components that help the brain behave normally. You undergo the operation and feel fine. You are still *you*. Or we can imagine attaching your brain to a wire that attaches to a warehouse filled with mechanical parts that work together to grind out an answer to a math problem or that preserves some memory of yours. Over many years, we may continue to replace portions of your brain until all we are left with is alternate representation of your entire mind. Using the bicycle-part analogy, your brain could be distributed over many locations, some in a warehouse in New Jersey and the rest in a warehouse in New York, interconnected by wires.

In principle, our minds could be hypostatized in the ripple patterns on a stream, in the movements of sand, or in the schooling behaviors of fish. If you could make a copy of your brain with the same structure but using different materials, the copy would think it was you. Yale

philosopher Nick Bostrom explores this line of thinking by starting
with a single brain cell:

> The brain cell is a physical object with certain characteristics. If
> we come to fully understand these characteristics and learn to
> replicate them electronically, then surely our electronic brain cell
> can perform the same function as an organic one. And if it can
> be done with a brain cell, why wouldn't the resulting system be
> just as conscious as a brain?[26]

The mind that emerges from the spinning gears of the bicycle brain
reminds me of the pointillist paintings of French artist Georges Seurat.
Up close, all we see are strokes or dots of contrasting pigments. Step back,
and the points blend together to form new colors, forms, and shimmering
lights to create totally new realities. Similarly, the bicycle brain is nothing
but gears and chains, but taken as a whole, the mind emerges.

For those of you who find it hard to believe that something as
mysterious as conscious thought could ever arise in a bicycle brain,
computer, or system of rubber bands and sticks, the discussion
becomes even more complex when you realize that our voluntary
motions are not even initiated by our conscious minds! Brain signals
relating to our acts occur *before* we have a conscious intent to do
something, such as deciding to push a button. If I had placed an elec-
trode in your brain, I could often detect when you were about to do
something before you knew. The average person believes that the
movement of his hand is initiated by his willpower. You think that
your psychological decision process is the trigger and prime cause of
that movement. However, 0.8 seconds before you consciously
decided to move your hands, there was an electrical signal in your
brain called a readiness potential. In other words, at the exact instant
you *consciously* decided to move your hands, the actions had been
already determined almost a second before. What would this startling property mean for simulated minds?

Debates continue to rage about the prerequisites for consciousness. In John Brockman's *What We Believe But Cannot Prove*, philosopher Daniel Dennett suggests that acquiring a human language (oral or sign language) is a necessary precondition for consciousness. On the other hand, psychologist Alison Gopnik believes that babies and young children are actually *more* conscious and more vividly aware of their external and internal life than adults are. After all, children have substantially more neural connections than adults. Psychologist Robert Provine suggests that humans really aren't even conscious most of the time. He writes, "Since we are not conscious of our state of unconsciousness, we vastly overestimate the amount of time that we are aware of our actions, whatever their cause. . . . Instead of wondering whether other animals are conscious. . . should we be wondering whether our behavior is under no more conscious control than theirs?" Neuropsychologist Stanislas Dehaene suggests that our minds are different from other animal minds due to certain neurons in the anterior cingulate region of our brains that send connections throughout the brain. This long-distance connectivity enables a "unique and flexible" communication between distant brain areas. If humans are indeed special because of enhanced information exchange between our various subbrains, perhaps one day we can enhance human potential by making appropriate increases in the connectivity.

Why Simulate?

Nick Bostrom suggests that simulated humans might be first created as simulations of historical individuals who are conscious and who believe themselves to be biological humans living in our past. Perhaps, individuals in advanced civilizations will simulate new realities to:

1. Rerun historical events with slight changes to observe the results
2. Produce great artworks akin to ballets or plays
3. Create lucrative and exciting tourist attractions

4. Explore new realms of thought and mind

5. Transcend space and time

6. Observe and partake in new sexual adventures

7. Perform illegal and sadistic acts

8. Find true love

9. Predict the future

10. Escape the present

11. Seek adventure and mystery

12. Create new sports

13. Solve problems

14. Meet God

Today, people enjoy playing computer and board games like The Sims, Civilization, Age of Empires, Diplomacy, and Monopoly for the same kind of escapism or feeling of control that they would experience in the future when tinkering with a detailed simulation of conscious beings and their artificial worlds.

As a motivation for option 1 (rerunning historical events), historians may wish to better understand why history happened as it did, or why certain trends are common in history. They may be able to learn something about the past and present by running several historical simulations and playing "What if?" games. In practice, these simulations would be difficult to run because we may not have sufficient information about motivations and people in the past to perform accurate simulations. Another challenge is the butterfly effect in history. As we discussed in the Introduction, much of history depends on contingencies—seemingly unexpected and insignificant factors that change the course of events. Perhaps simulations can be run to better understand which aspects of history are based on contingencies and which are based on large trends that are nearly unstoppable.

Wouldn't it be fascinating to rerun history to see the effects of inventions or discoveries that we force to never happen or suddenly appear? For example, imagine that the year is 1900 and a modern personal computer is given to scientists at Harvard University. It runs Microsoft Windows and comes with manuals. Assume that the system also comes with a power source that will allow it to run trouble-free for a year. How would humanity be affected from both a scientific and sociological standpoint? Would the concept of programming languages take the academic world by storm? Would scientists be just as interested in the computer packaging (plastic bags, bubble-wrap, corrugated cardboard shipping materials, cable ties) with a subsequent major impact on humanity? Would we have had computer-targeted field guns in World War I or computer navigation equipment available in World War II?

Computational historians in the year 2100 might be able to better understand why certain historical phenomena appear cyclically, why certain social movements grow, and why mass hysteria and fads take place. Economists and marketing people might run vast simulations to predict how mutual funds will do and how products might sell. Future theologians might even run and rerun certain biblical, ethical, or spiritual scenarios.

Regarding the difficulty of rerunning historical events with small changes, perhaps the best we can ever do is to create certain starting conditions and sets of beliefs using actors and simulations, and rerun rough historical approximations to see what may happen. It may be easier to start with the present, for which we have greater information, and simulate a number of futures to determine which may yield the greatest happiness.

For many of the options in our previous list of motivations for simulating reality, it is unnecessary to populate the synthetic worlds completely with conscious people. Clearly, an opera, play, ballet, or ball game might be beautiful, exciting, and novel with merely an 80 percent to 20 percent ratio of unconscious to conscious. In option 7, a truly sadistic person may require a fully conscious or even hyperconscious simulated person, for example, someone who can feel pain more than an ordinary person. Hopefully, strict laws would limit the number of such simulations. Similar simulations can be used to answer the question about why so much suffering takes place—perhaps it's not God who is allowing suffering but the simulator (an advanced being capable of performing a reality simulation) to whom God gives free will.

Option 8 probably requires the actual person to immerse himself completely in the simulated world in which he encounters his conscious or hyperconscious lover. Those options that involve predicting the future may be useful for those who want to avoid war, famine, or drought in the real world by running simulations and then offering advice to policy makers. Similarly, people may run simulations to pick stocks and increase wealth—if concepts of monetary wealth have any significance for a future civilization on the brink of posthumanism.

If we wish to solve problems in our simulations, imagine what it would be like to create a modified temporary clone of yourself with a kind of autism that allowed the clone to concentrate on problems more intensely than you while still possessing many of your traits and memories. Once your autistic simulacrum found the solution, the simulation is stopped and stored until needed again. Today, some scientists speculate that autism is caused by a lack of cooperation between different brain areas or by oddities in the organization and number of neurons in certain brain regions that may lead to connectivity problems. These deficits, which normally prevent a human from integrating well with other people, may not matter to you or your autistic simulacrum who is engineered to efficiently solve certain problems. If you want to "meet God" (option 14), certain brain states may be simulated that encourage hyperreligiosity as an aid to communing with God—and to seek transcendence.

Zombies, Peas, and Carrots

Crowds in a virtual city could easily be simulated without having to make the people conscious. They could be replicas of a basic theme, with different clothes, hair, and walking styles. Just as a laugh track machine is used to create an ambiance for TV shows, a simulated reality track can be used to create sounds and moods. Professor Robin Hanson, author of "How to Live in a Simulation," notes that it would be easy to simulate fake crowds by having all the "people" repeatedly mumble "peas and carrots" as do extras in restaurant scenes in movies to create a background din and realistic movement of lips. For increased complexity, half of the "extras" in the simulation can repeat "peas and carrots" while the other half says "ugamumble, ugamumble." According to gossip on the Web, this particular combination of "peas and carrots" and "ugamumbles" produces sounds surprisingly like background conversations. Good crowd sounds for radio shows also require a lot of indistinct mumbling interspersed with a few real words. American radio dramatists call this mumbling "walla," while in the UK it's called "rhubarb," and in Germany, "rhabarber."

Future scientists and artists will go far beyond sounds to create technologies that allow us to transcend our real world and get glimpses of beautiful synthetic ones. Futuristic "novelists" will capture awesome hyperrealities with evocative smells, sights, sounds, and a thousand other trans-sensorial elements to tickle the imagination. Imagine mystical scenes that seem utterly real, except perhaps for the presence of a shimmering velvet sky, silver crustaceans playing along the shoreline, bananas that sing, or sentient lichen atop little hobbit homes scattered across a verdant valley. Today, ordinary writers conjure mood and a sense of reality with their wonderful textual descriptions. The following are several snapshots from today's novels that I would like to experience in a Matrix, even if just for a few minutes. I've been collecting these favorite scenes for years:

> Tonight, the hay in the fields is already brittle with frost, especially to the west of Fox Hill, where the pastures shine like stars. In October, darkness begins to settle by four-thirty and although the leaves have turned scarlet and gold, in the dark everything is a shadow of itself, gray with a purple edge. At this time of year,

these woods are best avoided, or so the local boys say. (Alice Hoffman, *Here on Earth*)

They all lived in East Buckingham, just west of downtown, a neighborhood of cramped corner stores, small playgrounds, and butcher shops where meat, still pink with blood, hung in the windows. The bars had Irish names and Dodge Darts by the curbs. Women wore handkerchiefs tied off at the backs of their skulls and carried mock leather snap purses for their cigarettes. (Dennis Lehane, *Mystic River*)

The cascade of iridescence blew like musical notes of light; everyone was breathing out good wishes, frivolous and loving, evanescent and silly. . . . Slanted rays shone through a darkening swirl in the sky, rococo castles and feathers and mare's tails in the clouds. Someone was pulling out all the celestial stops. It was so beautiful that Mattie could hardly bear it. (Anne Lamott, *Blue Shoe*)

A mile more: of chastising thorns, burrs and briers that catch at our clothes; of rusty pine needles brilliant with gaudy fungus and molted feathers. Here, there, a flash, a flutter, an ecstasy of shrillings remind us that not all the birds have flown south. Always, the path unwinds through lemony sun pools and pitch-black vine tunnels. Another creek to cross: a disturbed armada of speckled trout froths the water round us, and frogs the size of plates practice belly flops; beaver workmen are building a dam. (Truman Capote, "A Christmas Memory")

In *The Matrix* movie series, we are told that UFOs, angels, ghosts, aliens, vampires, and werewolves are a result of an improperly functioning computer program. We can stretch our imaginations and speculate that all the psychiatric disorders in our world are the results of an improperly running simulation. Uncommon psychiatric disorders might be the result of one or two improperly coded lines of software that result in the range of very real psychiatric problems encountered today: the illusion of intermetamorphosis, delusional parasitosis, seeing doppelgängers, bipolar disorder, Charles Bonnet syndrome, Capgras syndrome, Fregoli syndrome, asomatognosia, prosopagnosia, redu-plicative paramnesia, the syndrome of subjective doubles, Cotard's syn-drome, and Ekbom's syndrome. Mysterious sightings of doppelgängers result from software copies. Already, sensitive people who have watched *The Matrix* report that they have direct evidence that we are truly living in a Matrix. One person writing to Nick Bostrom said that he could see flickering pixels when looking into his bathroom mirror![27]

Cosmic Reality Onion

All of this speculation about future possibilities of Matrix creation assumes that our civilization will survive long enough to embark on this posthuman adventure. It also assumes that we will be *interested* in trying to simulate complex realities. If we succeed at creating such a simulation, it's possible that the simulated civilization will someday also become posthuman (i.e., acquire the capability to simulate other humans). At some point, dozens of simulated civilizations within civi-lizations may exist like layers within layers of a cosmic onion. If the reality onion is designated by $\mathcal{N}_n$, with n being the reality level, we may designate actual reality as $\mathcal{N}_0$. The first simulated reality is $\mathcal{N}_{-1}$, the next $\mathcal{N}_{-2}$, and so forth. What can we say about a civilization as n decreases? The further away a civilization is from actual reality, the more sensitive it is to destruction because civilizations with larger values of n can just turn off a switch to destroy the realities beneath it. This may imply that the probability of existence of a reality level is directly related to the value of the level number.

Although it's possible to imagine how an $\mathcal{N}_n$ civilization might be able to interact and send messages to an $\mathcal{N}_{n-1}$ civilization, will ways exist by which we can skip levels, so for example an $\mathcal{N}_7$ can chat with an $\mathcal{N}_{20}$? Could an $\mathcal{N}_n$ civilization pretend to be Gods to an $\mathcal{N}_{n-1}$ and offer rewards and punishments in an afterlife for the $\mathcal{N}_{n\text{-}1}$? As Nick Bostrom points out, "an afterlife would be a real possibility, as would reincarnation." Would it be possible for $\mathcal{N}_n$ beings to reward $\mathcal{N}_{n\text{-}1}$ righteous beings by transporting them from the $\mathcal{N}_{n\text{-}1}$ universe to the $\mathcal{N}_n$ universe upon death? Perhaps the Christian Rapture whereby good Christians are lifted into heaven before the End of Days is easily accommodated in the onion universe. Bostrom wonders how would we know we are living in a layer of the onion:

> If we are in a simulation, is it possible that we could know that for certain? If the simulators don't want us to find out, we probably never will. But if they choose to reveal themselves, they could certainly do so. Maybe a window informing you of the fact would pop up in front of you, or maybe they would "upload" you into their world. Another event that would let us conclude with a very high degree of confidence that we are in a simulation is if we ever reach the point where we are about to switch on our own simulations.[28]

Imagine what a strange world we'd live in if the various $\mathcal{N}_n$ civilizations could interact. If only adjacent civilizations could communicate, the $\mathcal{N}_n$ onion would still form a kind of continuum; that is, even though individuals like you and me might not be able to observe the universes far removed, we could send them a message via a chain of $\mathcal{N}_n$ worlds that would relay the message from one $\mathcal{N}_n$ world to the next $\mathcal{N}_{n\text{-}1}$ world. Of course, it is hard to be sure if someone too far away to have direct communication is really there after all. And when a message got to the recipient, it might be distorted, and the $\mathcal{N}_n$ being might not believe that he had received a message from a creature he could not see, a creature that allegedly lived and had its being in some parallel world. Any response from the $\mathcal{N}_{n\text{-}1}$ worlds that traveled back up through the reality layers, distorted by the reverse chain of messengers, might seem to come from nowhere. And, maybe, sometimes, it would.

Perhaps if it were computationally expensive to simulate fully conscious people, we could still make realistic onion worlds with precisely ten thousand conscious people replicated millions of times. If the world were sufficiently big, then replicas might have little chance of meeting one another in their daily interactions.

In the movie *The Matrix*, evil machines in our N_0 world enslave us by sticking our minds in an N_1 world that they can rule. However, if it turned out that you and I were simulations in our own universe, perhaps there need be no evil aliens or machines behind the cosmic curtain. Nick Bostrom writes, "If the unfortunate reality is that we are simulations of some post-human civilization, then we are arguably better off than the inhabitants of the Matrix. Rather than being held captive by a malevolent AI in order to power their civilization, we have been created out of software as part of a scientist's research project. Or perhaps created by a post-human teenage girl for her science homework. Nevertheless, we're better off than inhabitants of the Matrix. Aren't we?"[29]

If we live in a cosmic onion, an evolutionary selection factor may exist for goodness and peacefulness in the onion multiverse. Only those civilizations that do not annihilate themselves through war will have developed sufficiently far to simulate new worlds. Additionally, in the $N_{n-1}, N_{n-2}, N_{n-3} \ldots$ synthetic subworlds, only those that are peaceful and advanced will survive to spawn new worlds.

Who might be simulating us? There are several possibilities:

1. Advanced humans
2. An alien race who has observed our radio transmissions and was able to resurrect us by analyzing our brain waves, DNA, and other biological sequences that exist in various databanks
3. Beings in parallel universes
4. Sentient computers
5. Other simulations
6. God

• • •

If we do live in a simulation, one can wonder again why there is so much pain and suffering? Are N_n beings sadists and inflicting pain on their N_{n-1} children? One answer is that to establish an unpredictable, creative, and infinitely evolving universe, N_n beings cannot directly reveal their existence to the N_{n-1} creatures. If the N_n superbeings did this, N_{n-1} intelligent life would be robbed of its independence. The N_{n-1} people would know too much. To stimulate creativity, the N_{n-1} beings must always search for answers on their own, with perhaps occasional hints from above. Why is life so painful? It is only in a crucible of competition that N_{n-1} intelligence, compassion, and art can flower.

Bits, Brains, and the Transreality Diaspora

Is there a limit to the complexity of a Matrix-like simulated world? According to Seth Lloyd of the Massachusetts Institute of Technology, if the visible universe were turned into a computer, it could process about 10^{120} bits of information in the age of the universe (13.7 billion years.) This defines the maximum available processing power for an advanced civilization (assuming that it does not have access to parallel universes). Any calculation that requires more than 10^{120} bits could not be undertaken because the computer could not carry this out in the time available.[30]

If we focus on a 1 kilogram computer, Lloyd calculates an upper bound of 5×10^{50} logical operations per second carried out on 10^{31} bits. The amount of computing power needed to simulate a human mind is estimated to be 10^{14} to 10^{17} operations per second for the entire brain.[31]

According to Nick Bostrom, a planetary mass computer performs 10^{42} operations per second using known nanotechnological designs. He says that "a single such computer could simulate the entire mental history of humankind by using less than one millionth of its processing power for one second. A posthuman civilization may eventually build an astronomical number of such computers."[32]

Other researchers have tackled similar problems from a different perspective. In 2004, Lawrence M. Krauss and Glenn D. Starkman of Case Western Reserve University, submitted "Universal Limits on Computation" to *Physical Review Letters*. They showed that the universe's *expansion* places limits on computation because it is not possible to transmit or receive information beyond the so-called global event-horizon in an accelerating, expanding universe. In particular, they determined how far an observer could travel in such a universe and still be able to transmit energy back to Earth. Their calculations show that the total number of computer bits that could be processed in the *future* would be less than 1.35×10^{120}. Their calculations also considered that the acceleration causes space to emit a form of energy known as de Sitter radiation, which would drain energy from regions of the universe and thus also diminish the computer resources that could be used for computation.[33]

If advanced beings find it very difficult to simulate complex worlds, they may find it easier to simulate less detailed virtual realities and a set of memories to support these realities. For example, someday it will be possible to simulate a visit to the Middle Ages and, to make the experience realistic, we may wish to ensure that you believe yourself to actually be in the Middle Ages. False memories may be implanted, temporarily overriding your real memories. This should be easy to do in the future—given that we can already coax the mind to create richly detailed virtual worlds filled with ornate palaces and other beings through the use of the drug DMT. In other words, the brains of people who take DMT appear to access a treasure chest of images and experience that typically include jeweled cities and temples, angelic beings, feline shapes, serpents, and precious, shiny metals. When we understand the brain better, we will be able to safely generate more controlled visions.

Our brains are also capable of simulating complex worlds when we dream. For example, after I watched a movie about people on a coastal town during the time of the Renaissance, I was "transported" there later that night while in a dream. The mental simulation of the

Renaissance did not have to be perfect, and I'm sure that there were myriad flaws. However, during that dream I *believed* I was in the Renaissance. If we understood the nature of how the mind induces the conviction of reality, even when strange, nonphysical events happen in the dreams, we could use this knowledge to ensure that your simulated trip to the Middle Ages seemed utterly real, even if the simulation was imperfect. As we discussed, it will be easy to create seemingly realistic virtual realities because we don't have to be perfect or even good with respect to the accuracy of our simulations in order to make them seem real. After all, our nightly dreams usually seem quite real even if upon awakening we realize that logical or structural inconsistencies existed in the dream.

Various brain regions—such as the medial prefrontal cortex, precuneus, and anterior insula—participate in the mind's sense of self. In the future, for the purposes of enhanced pleasure or mind expansion, we may temporarily adjust the functioning of the brain's "self-network" to literally become someone else in our virtual parallel dream lives and then later absorb their memories as we reintegrate. Current research has shown that people with abnormal activity in these brain regions may suddenly become fascinated with entirely new hobbies and become, for example, obsessed with painting, photography, or the collection of stuffed animals. They may change their political affiliations. Their senses of self become putty, ready to be molded and remolded again.

In the future, for each of your own real lives, you will personally create ten simulated lives. Your day job is a computer programmer for IBM. However, after work, you'll be a knight with shining armor in the Middle Ages, attending lavish banquets, and smiling at wandering minstrels and beautiful princesses. The next night, you'll be in the Renaissance, living in your home on the Amalfi coast of Italy, enjoying a dinner of plover, pigeon, and heron.

If this ratio of one real life to ten simulated lives turned out to be representative of human experience, this means that right now, you only have a one in ten chance of being alive on the actual date of today!

Our desire for entertaining virtual realities is increasing. As our understanding of the human brain also accelerates, we will create both imagined realities and a set of memories to support these simulacrums. Due to rise of synthetic worlds in online games, we are already seeing the tip of the iceberg where the distinctions between virtual reality and real reality blur or don't matter. Many friends spend hours in online virtual environments. Transreality economies have evolved in which we pay real money for virtual goods and virtual money for real goods. The Transreality Diaspora has already begun.

Process Physics and the Birth of Reality

Today, scientists are actively simulating and building realities using random numbers and mathematics in the modern field of Process Physics. Space, time, and all the objects in the universe are no more than a fizz in a mathematical soda of randomness, according to Reginald Cahill and Christopher Klinger of Flinders University in Adelaide. Whereas conventional physicists start with the idea that "objects" exist and are "real," Processes Physicists find it more satisfying to do away with this unnecessary layer of assumption.[34]

Cahill and Klinger begin by building realities from "pseudo-objects" that have no intrinsic existence—they are defined "only by how strongly they connect with each other, and ultimately they disappear from the model. They are mere scaffolding."[35] Next, the scientists add a little randomness and let the mixture percolate using simple mathematical rules. The system gradually evolves inside a computer. To define and track the pseudo-object connections, they use a 2-D grid of numbers in a "Reality Matrix." For example, pseudo-object 1 may have a connection strength with object 2 of S_{12}, and it has connection strength with object 3 of S_{13}, and so on.

To kick-start the formation of a new reality, the physicists fill the Reality Matrix with numbers that are very close to zero. The matrix is repeatedly transformed by a matrix equation that adds random noise, along with a nonlinear term involving the inverse of the original matrix. As the process is repeated, most of the elements in the Reality Matrix hover close to a value of zero, but some numbers suddenly become large. Startling structures begin to emerge. At this point in the evolution of the system, the reality scaffold can be visualized as branches drawn between pseudo-objects having strong connections. The scaffold evolves and grows as the process is repeated, and smaller branches connect to others. The various "trees" branch randomly, but begin to exhibit remarkable properties and structures:

> If you take one pseudo-object and count its nearest neighbors in the tree, second nearest neighbors, and so on, the numbers go up in proportion to the square of the number of steps away. This is exactly what you would get for points arranged uniformly throughout three-dimensional space. So something like our space assembles itself out of complete randomness. "It's downright creepy," says Cahill. Cahill and Klinger call the trees "gebits," because they act like bits of geometry.[36]

Although the connections between pseudo-objects decay, the connections are created faster than they decay. Eventually, the number of gebits increases exponentially, corresponding to a space that undergoes an accelerating expansion, just as space does in our universe. While some structures disappear, others become more elaborate. Something like matter emerges. "Topological defects" form in the fabric of connections—pairs of gebits that are far apart by most routes, but have other shorter links. They are like wrinkles in the fabric of space. Cahill and Klinger believe that real objects consist of these defects—as described by the wave functions of quantum theory—because they have a special property shared by quantum entities: nonlocality and a spooky form of entanglement or connectivity even for two faraway objects. Process physics is currently also being used to provide explanations for a myriad of phenomena including inertia, time dilation effects, gravity, black holes, event horizons, and the arrow of time![37]

You can read much more about Process Physics on the Web and how the pseudo-objects in our real world become hidden from view during the process of generating reality. According to Cahill and Klinger, "All we can measure is what emerges, and even though gebits are continually being created and destroyed, what emerges is smooth 3-D space. Creating reality in this way is like pulling yourself up by your bootstraps, throwing away the bootstraps and still managing to stay suspended in mid-air."[38]

I know that this discussion of Process Physics seems totally theoretical, but Cahill believes that one day we may discover that matter and the laws of physics emerge out of nothing but interconnected gebits. Interestingly, the continual generation of laws, events, and "matter" by the invisible scaffolding leads to a breakdown of cause and effect as reality is driven by a roulette machine at its deepest levels.

In short, process physics is a model of reality that replaces general relativity and unifies it with quantum theory. This system has no prior objects or laws, and evolves using an iterative system.[39] Quantum phenomena are caused by "fractal topological defects embedded in and forming a growing three-dimensional fractal process-space." "If we are right," Cahill says, "we may have produced the first new comprehensive mind-set for explaining reality for 2500 years."[40]

Their matrix has become *The Matrix*.

Quantum Resurrection and Immortality

All this talk of universes springing from random fluctuations or from supertechnologies rather than through the beneficent hand of God may make some readers nervous. If God did not create the universes, is there a hope of an afterlife filled with bliss? The good news is that we can at least be relatively sure that all of us will be resurrected.

Many astrophysicists today suggest that our universe will continue to expand forever and particles will become increasingly sparse. This seems like a sad and boring end, doesn't it? However, even in this empty universe, quantum mechanics tells us that residual energy fields will have random fluctuations. Particles will spring out of the vacuum as if out of nowhere. Usually, this activity is small and large fluctuations are rare. But particles *do* emerge, and given a long amount of time, something big is bound to appear, for example, a hydrogen or helium atom, or even an entire organic molecule like ethylene, $H_2C=CH_2$. This may be unimpressive to you, but if our future is infinite, we can wait a long time, and almost anything could pop into existence. Most of the gunk that emerges will be a hideous, amorphous mess, but every now and then, a tiny number of elephants, planets, people, or a Jupiter-sized brain made from gold will emerge. Given an infinite amount of time, you will reappear, according to physicist Katherine Freese of the University of Michigan.[41] Random fluctuations could even lead to a new big bang, but we'd have to wait a long time for that, on the order of 10 to the power of 10^{56} years.[42]

Quantum resurrection awaits all of us. Be happy.

• • •

And if the idea of waiting a long time for your quantum resurrection seems a little too boring for you—given your exciting and fast-paced life today—maybe you can derive more pleasure from the idea of "quantum immortality." To understand this concept, we must first understand the many-worlds interpretation of quantum mechanics, proposed in 1957 by physicist Hugh Everett III. This controversial theory posits that the universe at every instant branches into countless parallel worlds. More specifically, the theory suggests that whenever the universe is confronted by a choice of paths at the quantum level, it actually follows all possibilities, splitting into multiple universes. In the many-worlds theory, there is a vast number of universes, and if the theory is true, then all kinds of strange worlds exist. For example, in some parallel worlds, World War I never started because Archduke

Franz Ferdinand, the heir to the Austrian throne, slept late and was not assassinated on June 28, 1914.

According to proponents of quantum immortality, the many-worlds interpretation of quantum mechanics implies that a conscious being can live forever. Your heart disease will not kill you. Your fatal bicycle accident a year from now will never take place. The theory also means that terrorists continue to exist, even after their backpacks explode in crowded shopping malls. The strange logic for quantum immortality becomes clear in the following paragraph.

Suppose you are on an electric chair, which your executioner is about to turn on and send a jolt of 2,500 volts through your body. In almost all parallel universes, the electric chair will kill you. However, there is a small set of alternate universes in which you somehow survive—for example a rat may bite into some electrical wiring at the instant the executioner pulls the switch. Or the President may grant you a pardon. Or all the people in the execution room decide that capital punishment is immoral. The idea behind quantum immortality is that you are alive in, and thus able to experience, one of the universes in which the electric chair malfunctions or you are set free, even though these universes form a small subset of the possible universes. In this way, you would appear, from your own point of view, to be living forever.

We can conceive of a thought experiment for researching the many-worlds interpretation of quantum mechanics. Imagine that you sit in your basement next to a hammer that is triggered or not triggered based on the decay of a radioactive atom. A few minutes ago, you kissed your spouse good night. A Mozart sonata now plays in the background. Lavender incense fills the air. Take a deep breath.

With each run of the experiment, a 50-50 chance exists that the hammer will smash your skull, and you will die. If the many-worlds interpretation is correct, then each time you conduct the experiment in your basement, you will be split into one universe in which the hammer smashes and kills you and another universe in which the hammer does not move. Perform the experiment for a thousand days, carefully making a check mark in a notebook for each day that you survived. After many such attempts at quantum suicide, you may find yourself startled to be happy and still alive. In the universe in which the hammer falls, you are dead and cease to exist. However, from the point of view of the living version of you, the hammer experiment will continue running and you will be alive, because at each branch in the multiverse, a version of you that survives exists. If the many-worlds interpretation is correct, you will slowly begin to notice that you never seem to die.

Although the concept of quantum immortality is highly speculative, I am not aware of any laws of physics that this violates. The many-worlds interpretation of quantum physics does not suggest that *any* event is possible, but it does suggest that every outcome that is possible will branch off from any given instant in time. Given the quantum mechanical underpinnings of our universe, which includes such strange phenomena as quantum tunneling and resurrection, it might actually be difficult to think of scenarios that are not *possible*. If we view reality as an ever-branching river, we may hope that one of the branches will carry you along for many millennia through a delightful, margaritaceous stream of time.

Let's agree to meet a thousand years from now and have a party.

$$\bullet \; \bullet \; \bullet$$

The test of quantum immortality that we have been discussing seems to involve a procedure that could take several months to implement. However, you should be able to do a rapid many-worlds interpretation (MWI) test using a Hocker MWI Roulette Wheel, named after inventor Michael Hocker, an IBM employee. The hypothetical device makes use of one million radioactive atoms, each with a mutually independent probability of decay of 0.5 in an interval of one second. The device is attached to high-voltage equipment that will kill you if triggered by the decay of an atom. You simply flick the switch, and one million miniature experiments take place in a second. According to MWI, in some universes the atoms won't decay during your experiment. In fact, the radioactive compound may appear to behave normally up to the moment you flick the switch on, and perhaps all decay stops for a second once you flip the switch. If MWI is correct, you'll be able to flick the switch off after a second, and you'll judge the experiment a success.

Indeed the many-worlds interpretation provides an experiment to ensure that you become wealthy. You can use the same Hocker MWI Roulette Wheel. Just follow these instructions: 1) Invest heavily in a stock; 2) Turn on the Hocker MWI roulette wheel so that if the stock price does not increase in a day, you are killed; 3) Repeat step 1. If the many-worlds interpretation is correct, you are guaranteed to be wealthy. The experiment in Quantum Stocks is perhaps a superior way to test the many-worlds interpretation than previous scenarios, because you become richer than Bill Gates in the process.

Jell-O and Consciousness

I'd like to leave you with another uplifting thought. If we return to our discussion of simulations, we may be able to get another glimpse of our own immortality. Hans Moravec, Research Professor in the Robotics Institute at Carnegie Mellon University, once said,

> A good simulation can be conscious, just like we are. In fact, in some ways, I look at ourselves as just a kind of simulation. We're a conscious being simulated on a bunch of neural hardware, and the conscious being is only found in an interpretation of things that go on in the neural hardware. It's not the actual chemical signals that are squirting around, it's a certain high-level interpretation of an aggregate of those signals, the only thing that makes consciousness different from other interpretations, like the value of a dollar bill.[43]

Now, consider a cube of lime Jell-O, a mile on an edge. Somewhere inside this cube of Jell-O are all your thoughts, all your memories. Why? The particles in the Jell-O are moving randomly due to thermal vibrations. Each state of motion of those particles can be mapped into a state of some simulation of your consciousness. Given the number of particles in the Jell-O, it is likely that your mind, or at least fleeting instances of your mind, is in there. Your mind could be spread diffusely through the Jell-O and present in the form of interacting particles, or your mind might be localized to one precise region of the Jell-O. To reiterate, if our thoughts and consciousness do not depend on the actual substances in our brains but rather on the structures, patterns, and relationships between parts, then the Jell-O could think and contain an approximation of you. It would think it was you. Given this line or reasoning, all of us are alive, and hopefully happy, in the Jell-O in refrigerators across America, in the stones of the great pyramids, in the gold bars at Fort Knox, and in the interstellar dust. The gold bars make us live forever. Jell-O makes us live forever. We all lead virtual lives in Jell-O. We are immortal.

Of course, the Jell-O particles will also code for innumerable other lives and thoughts, some of which will not be conducive to the continuation of "your" consciousness. However, "you" will be in the set of particles in which your experience continues. The trail of thoughts that connects one of your thoughts to the next would be one trail of "you" through the Jell-O. On the other hand, another bonus is that "you" would also be contained in countless other trails as well.

Similarly, the Jell-O need not be all in one place, but could be partitioned into little boxes separated by miles and connected by thin tubes of Jell-O. To understand this better, if technology becomes possible to separate your left and right brain hemispheres by a mile and to connect them by an artificial corpus callosum, you would still be you. Moreover, if we could split your brain into a thousand parts, properly connected by rapid communications channels, you would persist. You could continue this brain separation until you are countless cells scattered across the planet, and still retain your hopes, loves, and dreams. You'll fondly reminisce about your childhood, your long-haired Slovenian lover, and the time you studied Italian in Venice. Your disembodied and scattered mind will still yearn for the delicious tastes of Valrhona vintage chocolate bars made from Criollo cocoa beans flown in from the Palmira plantation in west Venezuela. You'll still wonder if there is a God, and your widely scattered brain modules will come together and dream. Your mind could still fall in love with another mind.

᳇ ᳇

The world's largest Jell-O, a 7,700-gallon watermelon-flavored pink Jell-O made by Paul Squires and Geof Ross, worth $14,000, was set at Roma Street Forum, Brisbane, Queensland, Australia on February 5, 1981 in a tank by Pool Fab.
                —*Guinness Book of World Records*, 1991 Edition

If we live in a simulated reality, we should expect occasional sudden glitches, small drifts in the supposed constants and laws of Nature over time, and a dawning realization that the flaws of Nature are as important as the laws of Nature for our understanding of true reality.
                –John Barrow, "Living in a Simulated Universe"[44]

As the Dalai Lama said, "You can consider the possibility of living forever in a computer as a positive karma. . . ." And remember, if you really want it, you need not die. And my mission, like any true spiritual guide, is to give you this desire to live forever by teaching you happiness.
                –Rael, *Yes to Human Cloning: Eternal Life Thanks to Science*

The Singularity is a common matter of discussion in transhumanist circles. There is no clear definition, but usually the Singularity is meant as a future time when societal, scientific and economic change is so fast we cannot even imagine what will happen from our present perspective, and when humanity will become posthumanity.

–Anders Sandberg, "The Singularity"[45]

Imagine a world in which generations of human beings come to believe that certain films were made by God or that specific software was coded by him. Imagine a future in which millions of our descendants murder each other over rival interpretations of *Star Wars* or Windows 98. Could anything–anything–be more ridiculous? And yet, this would be no more ridiculous than the [religious] world we are living in.

–Sam Harris, *The End of Faith*[46]

The claims of mystics are neurologically quite astute. No human being has ever experienced an objective world, or even a world at all. You are, at this moment, having a visionary experience. The world that you see and hear is nothing more than a modification of your consciousness, the physical status of which remains a mystery. Your nervous system sections the undifferentiated buzz of the universe into separate channels . . . [which] are like different spectra of light thrown forth by the prism of the brain. We really are such stuff as dreams are made of.

–Sam Harris, *The End of Faith*[47]

If you knew that you were a simulated person [created for a party guest at a simulated party], and you wanted to live as long as possible, you might want to discourage anyone from leaving the party. If the simulation might end early were the future guest to become bored, you might also want to make sure everyone had a good time. And your motivation to save for retirement, or to help the poor in Ethiopia, might be muted by realizing that in your simulation, you will never retire and there is no Ethiopia.

–Robin Hanson, "How To Live In A Simulation"

FIVE
JESUS AND THE FUTURE OF MIND-ALTERING DRUGS
🦎 🦎

In which we encounter highway mega-messiahs, biblical use of marijuana, psychedelic snails in the Christian basilica at Aquileia, "kaneh-bosem," Carl Sagan, Jesus in movies, brain-eating monsters, Roger Corman, *Attack of the Crab Monsters*, Nancy Sinatra, Jack Nicholson, LSD, the value of mind-altering drugs, the brain's own marijuana, endocannabinoids, MDMA (ecstasy), "Lucy in the Sky with Diamonds," DMT, ayahuasca, Ibogaine, Myron Stolaroff, Einstein and Google, state-specific sciences, and Charles Tart.

If the words "life, liberty, and the pursuit of happiness" don't include the right to experiment with your own consciousness, then the Declaration of Independence isn't worth the hemp it was written on.

−Terence McKenna[1]

Jesus, Drugs, and 1950s Movies

Mega-messiahs are sprouting along the highways of America. Perhaps someday, no matter where you go, a giant Jesus will be watching you. For example, the "World's Tallest Uncrucified Christ," resides in Eureka Springs, Arkansas. The statue is seven stories high and weighs two million pounds. Many enraptured visitors see Jesus' eyes move, but this is probably explained by a trick of the light. In 2004, a sixty-two-foot-high Jesus statue rose near an interstate highway north of Cincinnati. This stupendous savior is surrounded by large flea markets and traffic jams. The artist, James Lynch, rendered Jesus from fiberglass and Styrofoam and is famous for other huge sculptures such as Caesar Palace's big Neptune in Las Vegas.

All around the world, scholars and lay people contemplate Jesus—

his life, his divinity, and exactly what he was doing in the ancient Mideast to inspire, heal, and create a cusp—a mucronation, a singular point on the curve of civilization, beautifully illustrated in mathematics near (0,0) for $y = x^{2/3}$.

According to a recent *BBC News* report, Jesus Christ and his apostles may have used a cannabis-based anointing oil—prepared from the marijuana plant—to help cure people with various diseases and ailments.[2] In particular, several American researchers claim that the oils Jesus used contained a cannabis extract called "keneh-bosem." Researcher and author Chris Bennett says,

> The holy anointing oil, as described in the original Hebrew version of the recipe in Exodus, contained over six pounds of keneh-bosem—a substance identified. . . as cannabis, extracted into about six quarts of olive oil along with a variety of other fragrant herbs. The ancient anointed ones were literally drenched in this potent mixture.[3]

Bennett claims that cannabis helps relieve some of the symptoms of epilepsy as well as many other ailments for which Jesus and the disciples provided relief—such as skin diseases, eye problems, and menstrual complaints.

As far-fetched as this all sounds, we do know that mind-altering drugs have been used by many religions over the centuries, and that cannabis was widely cultivated throughout the ancient Middle East. As far back as the 1930s, a handful of scholars identified the Bible's "fragrant cane" component of holy anointing oil, used by the Levite priests, as flowering cannabis, a link since accepted by some Jewish authorities. Of course, this line of thinking is not without controversy, although I have seen long lists of botanists, etymologists, anthropologists, mythologists, and linguists who accept the possibility that the Biblical keneh-bosem is cannabis.[4]

Evidence also suggests that early Christians shared psychedelic mushrooms and the spiritual visions they produced—as depicted in artistic mosaics from AD 330 in the Christian basilica at Aquileia, where Italy borders on modern-day Slovenia. The basilica images depict baskets overflowing with mushrooms and snails. According to scholars, the monks ate the snails that fed on the hallucinogenic mushrooms, rather than eating the mushrooms directly, which can lead to unpleasant gastrointestinal side effects.[5] Carl Ruck, professor of classics at Boston University, also suggests that cannabis was used by biblical peoples for healing and other purposes:

> Ancient wines were always fortified, like the 'strong wine' of the Old Testament, with herbal additives: opium, datura, belladonna, mandrake and henbane. Common incenses, such as myrrh, ambergris and frankincense are psychotropic; the easy availability and long tradition of cannabis use would have seen it included in the mixtures. Modern medicine has looked into using cannabis as a pain reliever and in treating multiple sclerosis. It may well be that ancient people knew, or believed, that cannabis had healing power.[6]

Ruck and Bennett are convinced that Jesus used cannabis. The word Christ means "the anointed one," and Bennett contends that Christ was anointed with chrism, a cannabis-based oil, that may have fostered his spiritual visions. The cannabis-laced anointing oil allowed the priests and prophets to commune with Yahweh. As further evidence, residues of cannabis have been found in Egyptian and Judeaic vessels used in medicinal and spiritual contexts. In the Old Testament, Levite priests used the oil to receive "revelations of the Lord."

What controversies would erupt today if a movie featured Jesus using marijuana-laced oil? The portrayal of Jesus in films goes back to 1897, in the form of short passion plays of a few minutes' duration. The first feature-length film, *From the Manger to the Cross*, came out in

1912–it was filmed in the Mideast to give the scenes a greater sense of realism. A series of big movies were popular in the 1950s, and they avoided any direct portrayal of Jesus. For example, wide-screen epics of the Roman Empire were quite popular, and Jesus was seen on the periphery in films like *The Robe* and *Ben Hur*. Only his legs are visible during the crucifixion in *Quo Vadis, The Robe*, and *Ben Hur*. I still remember the heart-rending scene of Ben Hur, crossing the desert and nearly dying of thirst. He is given water by a young Nazarene, a carpenter's son who we later find to be Jesus.

From 1950s Movies to LSD and Marijuana

For years, any 1950s movie fascinated me if it portrayed life during the time of Jesus. However, now that many '50s films have become available in DVD format, I've moved from the '50s classics to the seemingly trivial '50s science-fiction movies, like *The Brain from Planet Arous*, because they tell us about what Americans were dreaming and thinking about at a very special time in the world–the atomic age, a time of great technological change and of the Cold War between the US and the former Soviet Union. Movies are the paring knives that reveal the flesh of our cultural conscious. Movie projectors shine a light on our attitudes and beliefs. Or as Ingmar Bergman said, "Film as dream, film as music. No art passes our conscience in the way film does, and goes directly to our feelings, deep down into the dark rooms of our souls."[7]

The '50s science-fiction movies usually employed numerous scientific words to give the movies an illusion of scientific realism. The films are littered with rocketry, aliens, oscilloscopes, gauges with moving arrows, monsters created from ordinary life forms by exposure to radioactivity, space exploration, and all kinds of scary threats that stemmed from the Cold War mentality and fear of communism. Fredric Jameson notes in *Signatures of the Visible*:

> Arguably, the golden age of the fifties science fiction film, with its pod people and brain-eating monsters, testified to a genuine

collective paranoia, that of the fantasies of the Cold War period. . . . The enemy within is then paradoxically marked by non-difference. "Communists" are people just like us save for the emptiness of the eyes and a certain automation which betray the appropriation of their bodies by alien forms.[8]

Just a few science-fiction movies were made in the 1940s, but more than 150 SF movies appeared in the 1950s. The year 1957 was particularly notable with *The Incredible Shrinking Man, Quatermass II, The Monolith Monsters, Attack of the Crab Monsters, Not of this Earth,* and *The Undead,* the last three directed by Roger Corman.

Aside from the *Brain from Planet Arous,* I admired *Invaders from Mars* (1953), featuring a young boy who discovers Martians hiding underground and led by a disembodied bulbous head in a glass sphere. *The War of the Worlds* (1953) featured aliens destroying cities from stingray-like space ships. Aliens with pulsating heads capture nuclear physicists in *This Island Earth* (1955). In the movie, the aliens need our uranium! In *Earth vs. the Flying Saucers* (1956) aliens destroy Washington, DC, the most important symbol of Western democracy. Other movies of the time include: *The Day the Earth Stood Still* (1951), *The Thing* (1951), *The Beast From 20,000 Fathoms* (1953), *It Came From Outer Space* (1953), *The Creature From the Black Lagoon* (1954), *Them* (1954), *Earth vs. the Flying Saucers* (1956), *The Day the World Ended* (1956), *It Conquered the World* (1956), *Creature with the Atom Brain* (1956), *The Invasion of the Body Snatchers* (1956), *I Was a Teenage Werewolf* (1957), *The Incredible Shrinking Man* (1957), *Not of This Earth* (1957), *The Fly* (1958), and *Attack of the Fifty Foot Woman* (1958).

Patrick Luciano notes in *Them or Us: Archetypal Interpretations of Fifties Alien Invasion Films,* "The proliferation of science fiction films is one of the most interesting developments in post-World War II film history. An estimated 500 film features and shorts made between 1948 and 1962 can be indexed under the broad heading of science fiction. One might argue convincingly that never in the history of motion pictures has any other genre developed and multiplied so rapidly in so brief a period."[9]

Of course, the skyrocketing number of SF movies in the 1950s was not only a response to the Communist threat. Some people were also concerned about the creeping uniformity in suburbia, environmental hazards like radiation, and our over-reliance on the deterrent philosophy of mutually assured destruction (MAD)—the military strategy in which use of nuclear weapons by one side would ensure the destruction of both the attacker and the defender.

One particular favorite '50s movie was Roger Corman's *Attack of the Crab Monsters*, which featured giant crabs mutated by atomic tests. On one level, it's a scary survivor story as the crabs assimilate the voices and intellects of their victims who are trapped on a remote Pacific island. Here is a snippet of the scintillating dialogue:

Dale: "That means that the crab can eat his victim's brain, absorbing his mind intact and working."

Karl: "It's as good a theory as any other to explain what's happened."

And the huge crabs themselves exhibit dialog that is not exactly Shakespearean:

Crab Monster: "So you have wounded me! I must grow a new claw, well and good, for I can do it in a day, but will you grow new lives when I have taken yours from you?"

On a deeper level, the movie poised humanity on the edge of an apocalypse—during a time that one age was ending to make room for a high-tech age in which humans manipulate nature in profound ways. The crabs were destroying our world but so were the humans with their newfound technical powers.

The movie starts with actual footage of atomic bomb blasts and mushroom clouds at the Eniwetok atoll in the central Pacific Ocean. Fifteen minutes into the movie, a sailor is mysteriously decapitated. Often featured in the film are the crabs' large, menacing claws along with the frightful clicking sounds they make. I recall that although the creatures look like giant crabs, their eyes were vaguely humanoid. Oh, those giant claws haunted me for much of my childhood!

The giant crabs looked very realistic given the limited special effects budget. They were built by a Hollywood firm using papier-mâché, and stretched about fifteen feet from side to side—held together with piano wires. To save money, Corman had the actor who played a naval officer also work inside the crab to make it move. Other crew members used support poles to help manipulate the monster. Much of the shooting was at the rocky Leo Carrillo State Beach in Malibu, California. Corman also tried to shoot underwater scenes with the hollow crab, but the creature insisted on floating, which required the filmmaker to weight it down with rocks and other objects.[10]

Crab Monsters was made in 1957 for $70,000, and grossed approximately $1 million, making it independent filmmaker Roger Corman's most profitable picture of the period. Corman, born in 1926, is famous today for his prolific output, and he represents an amazing motion picture success story. Having produced more than five hundred films, his influence on American culture obviously goes well beyond his *Crab Monsters*. Over the years, he has identified and fostered a number of talented producers, directors, writers, and actors including: Jack Nicholson, Francis Ford Coppola, Peter Fonda, Bruce Dern, Diane Ladd, Robert DeNiro, Martin Scorsese, Ron Howard, Charles Bronson, Joe Dante, and James Cameron.

Of his boyhood years, Corman writes in his autobiography *How I Made a Hundred Movies in Hollywood and Never Lost a Dime*:

My studies focused on sciences and math, but I read a great deal of literature as well, including Edgar Allan Poe's "The Fall of the

House of Usher," which undoubtedly made quite an impact. It was a class assignment, but I enjoyed it so much I asked my parents to buy the complete works of Poe for a birthday or Christmas gift. Who knew that twenty years later I would bring a half-dozen or so of those stories to the screen.[11]

The chameleonic Corman graduated from Beverly Hills High School, and in 1947 he received a bachelor's degree in engineering from Stanford University. He did postgraduate work in modern English literature at Oxford's Balliol College.

In 1953, Roger Corman sold his first screenplay, *Highway Dragnet*, and served as associate producer on the film. With the proceeds of the sale, he made *The Monster from the Ocean Floor* the following year with a budget of only $18,000. This was his first film as an independent producer. Corman turned out nine films in 1957—some of which were completed in two or three days!

Corman's films were usually box-office successes. Despite the limited special effects and occasional clunky dialogue, most of his films had serious moral and mystical implications. His science-fiction films often warned about the dangers of nuclear destruction. *The Day the World Ended* (1956) and *The Last Woman on Earth* (1960) dealt with the aftermath of nuclear holocausts. *Teenage Caveman* (1958) featured a "prehistoric" boy, whom the audience learns at the end of the film is actually living in a future devastated by nuclear war.

Always the chameleon, Corman traveled far beyond his science-fiction beginnings. For example, his first biker movie, *Wild Angels*, starred Peter Fonda and Nancy Sinatra. His movie *The Trip*, starring Jack Nicholson, began the psychedelic film craze of the late 1960s.[12] Over the years, he experimented with other genres including Westerns, horror flicks, thrillers, road movies, and drug and rock 'n' roll movies. In the 1960s, Corman purchased foreign-language films with good special effects and inserted segments featuring famous American performers to create "new" films.

LSD was the centerpiece of Roger Corman's 1967 movie *The Trip*, written by Jack Nicholson, with music performed by the psychedelic rock band, *The Electric Flag*. Corman has since revealed in interviews that he tried LSD under a doctor's supervision in order to prepare himself for the making of the film. He called his own LSD adventure "fantastic."[13]

The hero of *The Trip*, played by Peter Fonda, takes LSD for mystical purposes and embarks on a shattering odyssey of self-discovery. At the beginning of the trip, Fonda is in a state of bliss as he studies an orange and sees the universe within it. Corman recreated an LSD experience with colorful lighting and psychedelic music. Several of the visions in the trip are quite startling, including a castle with torturers and dwarves. Naked women and hooded women on horseback chase Fonda. Moiré patterns are projected over naked bodies. In the end, Fonda meets a woman, makes love to her by the ocean, and falls into a restful sleep. The next day, he is asked if the trip was worth it? He hesitates and says, "I suppose so."

Those three final words whipped a few critics into a frenzy because they seemed to endorse LSD experimentation—even though the movie opened with a disclaimer warning that LSD is dangerous.

The Mystery and Utility of Mind-Altering Drugs

In my book *Sex, Drugs, Einstein, and Elves*, I discussed how users of certain psychedelic drugs often felt that they were freed from time and became certain that their minds would be immortal and survive their deaths. Though mediation or the use of drugs, past, present, and future can lose their significance. Hypnosis can also cause time dilation (slowing), as can cannabis and LSD. On the other hand, heat appears to speed up the activity of a chemical timepiece in the brain. For example, fever can severely speed your perception of time, perhaps partly because it speeds chemical processes. Opium is notorious for its effect on time perception. The English writer Thomas De

Quincey reported that under the influence of opium he seemed to live as much as one hundred years in a single night. Another Englishman, J. Redwood Anderson, took hashish and said, "Time was so immensely lengthened that it practically ceased to exist." This reminds me of Tennyson's Lotus Land, "where it was always afternoon" or Samuel R. Delany's *Dhalgren* in which seconds sometimes lasted for hours.

Carl Sagan (1934–1996)—astronomer, exobiologist, writer, and TV star—achieved rock-star status with his creative output and energetic personality. He also used mind-altering drugs. Born to a Brooklyn Jewish family, Sagan became world famous for his popular science books, which discussed topics ranging from human evolution to the farthest reaches of the galaxy. His television series *Cosmos* won an Emmy and a Peabody Award and has been broadcast in sixty countries and seen by more than five hundred million people. The show focused on cosmology, astronomy, the origin of life, and humanity's place in the universe. His Pulitzer Prize–winning book *The Dragons of Eden* discussed anthropology, evolutionary biology, psychology, computer science, and human intelligence.

Sagan, like other amazing chameleons, was an avid user of marijuana. The drug improved his powers of lateral thinking and the ability to make novel mental connections. According to Keay Davidson, author of a definitive biography on Sagan, *Dragons of Eden* was written "under the inspiration of marijuana." Sagan clearly believed that pot enhanced his creativity, insights, and appreciation of art—and felt "convinced that there are genuine and valid levels of perception available with cannabis (and probably with other drugs) which are, through the defects of our society and our educational system, unavailable to us without such drugs."

Given this linkage between mind-altering drugs, such as marijuana, and chameleons, I would like to devote much of this chapter to drugs and the law. In fact, can mind-altering drugs ever be good for society? Imagine that you are dying of cancer—depressed, afraid, and in some pain. Would you be open to trying certain mind-altering drugs (e.g., MDMA, LSD, DMT, psilocybin, pot, or some new drug created for easing the transition to death), if you could be relatively sure that the drug would ease the pain, decrease your anxiety, and put you in a state of bliss, hopeful expectation, a feeling of being close to the divine and on the threshold of something wonderful? In short, would you choose to die while blissfully tripping? For the moment, let's assume that taking these drugs at a particular dosage would still allow you to communicate sufficiently well with loved ones and care givers, and that you would not be cut off from people and would be able to assist with your own bodily needs.

Society's nervousness about marijuana—particularly for medical use—is sometimes confusing. According to the editors of *Scientific American*, "outdated regulations and attitudes thwart legitimate research with marijuana. American biomedical researchers can more easily acquire and investigate cocaine."[14] Marijuana is classified as a Schedule 1 drug, alongside LSD and heroin, because it is potentially addictive and supposedly has no medical value. However, we know that the human brain naturally produces compounds, called endogenous cannabinoids, very similar to those found in marijuana. This natural endocannabinoid system in humans plays a role in the processing of pain, memory, anxiety, hunger, vomiting, neurodegeneration, and inflammation. We also know that marijuana has the potential to treat pain, reduce injury to nerves, and ameliorate nausea associated with chemotherapy.

Several mind-altering or psychedelic drugs may have value in treating anxiety and other ailments. For example, psychiatrist Francisco Moreno of the University of Arizona in Tucson has found that psilocybin may have the potential to safely treat obsessive-compulsive disorder. At the end of 2004, the Food and Drug Administration gave permission to a Harvard University plan to study the recreational drug ecstasy (also known as MDMA) as a treatment for anxiety in terminal care patients.[15]

At his South Carolina clinic, the psychiatrist Michael Mithoefer is conducting another study in which he gives MDMA to victims of sexual abuse with post-traumatic stress disorder in order to determine possible therapeutic value of MDMA. Mithoefer is preparing to extend the study to US soldiers traumatized by fighting in wars. Mithoefer's drug-assisted sessions last about seven hours each, during which clients recline and listen to music–though psychedelic songs like the Beatles' "Lucy in the Sky with Diamonds" are avoided. In fact, all music with lyrics is excluded from sessions so that people can create and imagine their *own* content.

Political prejudice seems to be denying people access to safe painkillers. For example, opioid drugs, such as morphine and Oxy-Contin, are among the most effective and safest painkillers, yet paranoia over their addictive properties has sometimes frightened doctors away. This means that patients are allowed to take risky drugs, even though the opioids are amazingly safe if not misused. "Entire government bureaucracies–from the US Drug Enforcement Administration to state polices and prosecutors–are devoted to demonizing opioids."[16] In fact, some doctors who are wary of government surveillance now avoid the prescription of narcotics like OxyContin, Percocet, Dilaudid, and Vicodin, causing patients to suffer needlessly. Nonsteroidal anti-inflammatory drugs like ibuprofen and aspirin, which can cause bleeding in the digestive tract, cause three times more deaths a year than OxyContin.

From time to time, scientists have become concerned about possible mental health risks associated with marijuana use. For example, some research has suggested a possibility that cannabis increases the risk of schizophrenia and psychosis in certain small groups of young people who have a particular kind of genetic disposition–namely people with two copies of a "bad" form of the gene known as "COMT." Even if this proves to be a valid association, it is unclear that policy for an entire nation should be based on a minority of people who may have this higher risk, and we clearly don't advocate illegalization based on health of minority populations when it comes to alcohol, tobacco, or numerous prescription and nonprescription drugs.[17]

Personally, I'm dubious of the suggested link between marijuana and schizophrenia, given the available evidence. As Graham Lawton in *New Scientist* says, "Despite a steep rise in cannabis use among Australian teenagers over the past 30 years, there had been no rise in the prevalence of schizophrenia."[18] In other words, if marijuana use increases, one would expect the occurrence of schizophrenia to increase, if the link between pot and schizophrenia were causal. Similarly, Lawton wonders if marijuana laws should be based on the benefit of a vulnerable minority. He notes that a minority of people is vulnerable to liver damage even if they drink a small amount of alcohol, but we haven't changed laws to accommodate this risk group.[19] In any case, more recent research suggests that the majority of people who had recently used cannabis already had schizotypal symptoms before using the drugs. These symptoms involved odd, magical beliefs and suggest that people with such traits are attracted to cannabis.[20] On the other hand, recent research shows that one of the chemical ingredients in marijuana, cannabidiol, actually has powerful antipsychotic effects. British researchers now plan to investigate cannabidiol as a treatment for schizophrenia![21]

Beyond Marijuana

What about the potential evils of other mind-altering drugs? Should these drugs be illegal to all? For example, John Halpern, associate director of substance abuse research at Harvard University's McLean Hospital, has studied members of the Native American Church, who are legally permitted to consume peyote. Halpern examined residents of a Navajo reservation in the southwest US who had taken peyote at least one hundred times, and he tested these people's IQ, memory, reading ability, and other functions. He found that church members had no cognitive impairment compared to control subjects and scored significantly better than recovering alcoholics who had been sober for over three months. Later studies indicate that moderate users of MDMA (ecstasy) had no major problems, and he hopes to test MDMA's psychological effectiveness to ease cancer patients' anxiety as they transition to death.[22]

Of course, the use of drugs like MDMA and other powerful mind-altering drugs has associated risks. People with a family history of depression may need to be particularly careful because MDMA might trigger depression. Ecstasy works by elevating levels of neurotransmitter serotonin, controlled by serotonin transporter genes. David Rubinsztein and colleagues at the University of Cambridge studied people with two copies of a shorter version of this gene and found that ecstasy users are much more likely than cannabis smokers and drug abstainers to suffer from clinical depression. If MDMA does become used in the clinical setting, perhaps clients should be screened for the kinds of serotonin transporter genes they have.[23]

Alex Polari de Alverga's *O Livro das Mirações* (*The Book of Visions*), describes the Santo Daime religion and community, located in the Amazonian rain forest.[24] Members of the religion drink ayahuasca as a shortcut to spiritual transcendence. Ayahuasca is a DMT-rich psychedelic brew from Amazon plants. After years of study, the Brazilian government decided that ayahuasca benefited both society and church members, and declared the drug legal. In fact, the government said that ayahuasca exerted an overwhelmingly positive influence on the lives of users, particularly when taken in a religious setting.

Charles Grob, a psychiatrist at the Harbor-UCLA Medical Center in Los Angeles, California, has studied ayahuasca in other Brazilian religions. Since 1987, it has been a legal sacrament for several churches in Brazil, the largest of which is União Do Vegetal—a blend of Christianity and nature worship. Grob found that church members who regularly took ayahuasca were on average physiologically and psychologically healthier than a control group of non-worshippers.[25] The church members also had more receptors for the neurotransmitter serotonin, which has been linked to lower rates of depression and other disorders.[26] Ayahuasca had also helped many church members overcome alcoholism and other self-destructive behaviors. Teenagers who grew up drinking the ayahuasca potion were less likely to engage in crime or substance abuse than control groups.

The West-African plant alkaloid, Ibogaine, may someday be useful for easing addiction and withdrawal symptoms. Howard Lotsof, famous for his research on Ibogaine, suggests that the psychedelic drug provides a cure for heroin and cocaine addiction. A former heroin addict himself, Lotsof first obtained Ibogaine from a drug researcher cleaning out his refrigerator. Lotsof found that the Ibogaine removed his own dependency on heroin and cocaine without the pain of withdrawal—a claim endorsed by other addicts who have tried it.

Two decades later, Lotsof patented the iboga molecule under the name Ibogaine for purposes of addiction treatment, but the FDA refused to approve it. Ibogaine was subsequently declared, along with LSD, an illegal Schedule 1 substance, with potential for abuse and no medical value—although it is legal in most of the world. The Food and Drug Administration (FDA) approved a clinical trial in 1993, but the National Institute on Drug Abuse (NIDA) decided not to fund it because of safety concerns. Nevertheless, thousands of patients use Ibogaine secretly and have achieved very good results.[27]

Physicians Jeffrey Kamlet and neuroscientist Deborah Mash said that for ninety days after treatment with Ibogaine, patients reported "feeling wonderful" and showed improvement in depression and reduced craving for drugs.[28] The liver converts Ibogaine into noribogaine, which fills opiate receptors for heroin or morphine. Noribogaine may also have antiaddictive value without the psychedelic effects of Ibogaine.

Difficulty for Research

Some of the most interesting early clinical experiments with LSD were performed by Myron Stolaroff, a Stanford-educated electrical engineer who spent his industrial career at the Ampex Corporation, a manufacturer of magnetic recording equipment. He had his first experience with LSD in 1956, at which point he said that LSD was the "most important discovery of mankind," and he came to believe that LSD provided humanity with a powerful tool for stimulating creativity. His life-altering experiences convinced him to devote the rest of his life to understanding the potential of psychedelic substances for psychological growth and for fostering innovation.

In 1961, he founded the International Foundation for Advanced Study in Menlo Park, California, where he conducted LSD and mescaline experiments for several years on hundreds of subjects. The FDA revoked all permits for research with psychedelics in 1965, but his work continued on a variety of mind-altering phenethylamine compounds until the Controlled Substance Analogue Act of 1986 was passed.

One of Stolaroff's early experiments involved dozens of volunteer engineers who were given LSD under constant observation. The subjects were instructed to take notes. A survey of the first 153 volunteers revealed that 83 percent of those who had taken LSD found that they had "lasting benefits from the experience."[29] Over 70 percent said that the LSD increased their ability to love and increased their self-esteem.

In another experiment, Stolaroff's foundation gathered together engineers to solve difficult problems while under the influence of LSD. Electrical engineers designed circuits, and Hewlett-Packard mechanical designers improved lighting designs. Even architects were creating plans for buildings while under the influence of LSD. John Markoff in *What the Dormouse Said* notes that many of the engineers and programmers who contributed to the birth of the personal computer are known to have consumed LSD. Perhaps counterculture did contribute to the birth of the personal computer. *New York Times* columnist Andrew Leonard observes:

Mr. Markoff makes clear his belief that computers, like psychedelic drugs, are tools for mind expansion, for revelation and personal discovery. And to anyone who has experienced a drug-induced epiphany, there may indeed be a cosmic hyperlink there: fire up your laptop, connect wirelessly to the Internet, search for your dreams with Google: the power and the glory of the computing universe that exists now was a sci-fi fantasy not very long ago, and yes, it does pulsate with a destabilizing, revelatory psychic power.[30]

I agree that Google is a vehicle for mystic transport with its endless panoply of topics that induce a pleasant neural overload. Writer Thomas Friedman is also in awe of Google and wonders, "Imagine if Einstein had Google and high-speed DSL, what he could have done during his lifetime on top of $E = mc^2$."[31]

Some of my engineer acquaintances have felt a kinship with Stolaroff's high-tech subjects who took LSD. For example, one colleague notes:

I was never able to write a decent song until I took LSD. I feel like I became very very open-minded as a result of taking the drug. It also seems like my ability to reason abstractly was expanded. My level of awareness of mathematical ideas was taken to a whole new level—I could look at problems from many angles and from meta-viewpoints. It also peeled away my subconscious so I could directly see my thoughts and understand them rather than having them in the background. These new insights and abilities persist to this day.[32]

Scientists find it very difficult to determine possible benefits of psychedelic drug use due to the extreme bureaucratic roadblocks faced when performing studies. For example, *Scientific American* recounts the story of one researcher who applied for a grant to study marijuana's potential medical benefits.[33] The National Institute on Drug Abuse (NIDA) rejected the grant. The same scientist reworded the grant to emphasize finding marijuana's negative effects, and NIDA funded the study instantly.[34]

Although there is evidence of the therapeutic value and limited toxicity of marijuana in the treatment of various afflictions, including glaucoma, migraine, epilepsy, multiple sclerosis, paraplegia, quadriplegia, the AIDS wasting syndrome, anxiety associated with impending death, or nausea and vomiting due to chemotherapy for cancer, the US government has refused to permit prescription sales of marijuana.[35] When Donald Abrams, of the University of California, San Francisco, sought permission to conduct a privately financed study comparing the effectiveness of inhaled marijuana with prescription cannabinol pills, for the treatment of weight loss associated with the AIDS wasting syndrome, the US government did not allow Abrams to obtain a legal supply of marijuana from Hortapharm, a company licensed by the government of the Netherlands to cultivate cannabis for botanical and pharmaceutical research. Abrams had to wait nine months to receive his letter of rejection from The National Institute on Drug Abuse (NIDA), which controlled the domestic supply of marijuana for clinical research![36]

Still, many people afflicted with diseases often go underground to obtain the drug that provides relief. As one example, Valerie Corral, cofounder and director of a medical marijuana collective in California, has found relief from her epileptic seizures through marijuana. Prior to smoking marijuana, she would have up to five grand mal seizures a day and paralyzing headaches, while living "under a waterfall of [prescription] pharmaceutical drunkenness." She found that she could "keep the seizures at a distance" through judicious marijuana use, and she "became convinced by the miracle of my own experience."[37]

Studies published in 2005 suggest that marijuana may someday help in the treatment of atherosclerosis, a disease associated with the accumulation of immune cells, lipids, and plaques in blood vessels. At low doses that do not produce psychotropic effects, THC (delta-9-tetrahydrocannabinol), a psychoactive compound in cannabis, can prevent blood vessels from developing atherosclerosis, at least in mice. Although THC is the best known active ingredient of marijuana, it is one of over sixty different cannabinoids that activate receptor molecules in the brain.

The Brain Is Its Own Drug

Cannabinoids are a group of chemicals found in marijuana that are responsible for the plant's peculiar effects on the human body. In addition to these molecules, similar chemicals called endogenous cannabinoids (endocannabinoids) are produced by our own bodies.

Cannabinoids produce their effects through interaction with microscopic cannabinoid receptors (CB1 and CB2) that are found in mammals, birds, fish, and reptiles. Our understanding of the cannabinoid receptor system is fairly recent. In 1988, researchers led by Allyn Howlett of St. Louis University School of Medicine, Missouri, discovered that cannabinoids bind to receptor molecules located on the surface of cells in the brain and nervous system, altering the way they function.[38]

Today, we know that the cannibinoid molecule binds to receptors throughout the nervous system, with high concentrations in the cerebral cortex, hippocampus, hypothalamus, cerebellum, basal ganglia, brain stem, spinal cord, and amygdala.[39] The brain's naturally occurring endocannabinoids play a key role in regulating other neurotransmitter signals in the brain. For example, when endocannabinoids bind to their brain receptors, they can sometimes block cells from releasing excitatory neurotransmitters. From a behavioral standpoint, the endocannabinoids can decrease the incidence of negative emotions and pain triggered by reminders of past experiences.[40] Low numbers of cannabinoid receptors or the faulty release of endogenous cannabinoids may trigger phobias, chronic pain, and post-traumatic stress syndrome.[41] Cannabinoids hold great promise for treatment of diseases like epilepsy, stroke, and multiple sclerosis because the cannabinoids reduce the severity of excitotoxicity, a common cause of neuron damage due to overexcitation.[42] Because the cannabinoid receptor is present in all vertebrate species tested, perhaps brains have made use of their own marijuana for millions of years. In fifteenth-century Iraq, marijuana was used to treat epilepsy. In the late 1800s and early 1900s, cannabis was available with a prescription and used to help with migraines and ulcers, and today marijuana is used in traditional medicine worldwide to alleviate pain, reduce muscle spasms, to reduce the number and severity of seizures, and to aid sleep.

Despite these positive uses, it is premature to think of cannabis as a wonder drug, because several compounds in cannabis, including cannabidiol and THC, interfere with natural signaling systems in the brain. The effect of swallowing or smoking cannabis can have variable results that depend on dosage as well as duration and timing of use. As pointed out by Helen Phillips in "Medical Cannabis Is a Blunt Tool" (*New Scientist*, July 29, 2006, p. 17), in the future, more useful therapeutics may actually involve amplifying the natural cannabinoid system, just as Prozac amplifies brain serotonin levels by inhibiting its reuptake by the brain.

Substitutes for Marijuana

I have a few friends who say that it's morally okay for people to take THC (tetrahydrocannabinol, the psychoactive component of marijuana) in pill form if prescribed by a doctor, but do not wish to permit patients to smoke marijuana, even if a doctor could prescribe this. Lawmakers also often suggest that only the pill form should be studied. On the other hand, smoking marijuana sometimes seems to provide benefit over the pill because it produces a more rapid increase in the blood level of the active ingredients, and some users claim that the pill just isn't as effective as smoking the cannabis plant. In 1992, synthetic THC was licensed for combating the nausea symptoms of AIDS, but as with multiple sclerosis patients, many AIDS patients found marijuana more effective.[43]

Several synthetic THC analogues are commercially available, such as nabilone and dronabinol.[44] Dronabinol stimulates appetite in AIDS patients. Other delivery vehicles are currently being investigated such as Sativex, an under-the-tongue cannabis spray for patients with multiple sclerosis. In 2005, Canada gave permission for it to be used for pain relief in adults with this disease. A 2004 study showed that Sativex provided significant pain relief for about 80 percent of people with MS.[45] In 2006, the FDA approved clinical trials of Sativex in America. Researchers are now planning studies for the possible benefits of standardized cannabis preparations for people afflicted with rheumatoid arthritis and heroin addiction.

In 2005, researchers at the University Hospital of Munich began the first clinical patient trial examining the efficacy of cannabis extracts as a treatment for Crohn's disease, a chronic inflammation of the intestine. Additionally, neuropsychologist Xia Zhang and a team of researchers based at the University of Saskatchewan in Saskatoon, Canada, are studying HU210, a synthetic drug that is about one-hundred times as powerful as THC. HU210 induces new brain cell growth, and Zhang suggests that this may potentially be therapeutic in reducing anxiety and depression. The following table lists several cannabinoid-based medicines, some of which are still undergoing clinical trials or are available only outside the US.

Cannabis Drug Potential Use

Sativex by GW Pharmaceuticals	Under-the-tongue spray to relieve multiple sclerosis pain.
Acomplia by Sanofi-Aventis	Raises "good cholesterol" levels. Improves insulin resistance in the obese. Potential use for alcohol withdrawal.
Cannabinor by Pharmos Corp.	Decreases postoperative pain when injected intravenously.
Nabilone available as Cesamet (capsule form) in Canada from Lilly	Prevents nausea and vomiting that may occur after treatment with cancer medicines.

| Dranabinol sold as Marinol by Solvay Pharmaceuticals | Prevents anorexia associated with weight loss in patients with AIDS, and nausea and vomiting associated with cancer chemotherapy. |

Making Nirvana and Pain Relief Illegal

When must a society make a drug illegal to use under all circumstances? Is it when a certain number of deaths or broken homes can be directly linked to the drug? Should we ban all mind-altering drugs, even when they may have medically beneficial uses? As we have seen, numerous pieces of evidence exist that suggest that hallucinogens can be effective against mental illness, including anxiety, post-traumatic stress disorder, alcoholism, and heroin addiction.[46]

In a 2006 study, Roland R. Griffiths of Johns Hopkins University School of Medicine and his colleagues tested psilocybin, the psychedelic agent from certain mushrooms, on thirty-six volunteers who had had never before taken hallucinogenic drugs. In follow-up interviews conducted two months later, two-thirds of the participants rated their psilocybin experience as among the most meaningful of their lives. They reported feelings of deep joy, unity with all things, a transcendence of time and space, an experience with immortality. Seventy-nine percent reported that the experience had moderately or greatly increased their overall sense of well-being or life satisfaction. What if we are able to invent safe drugs that help us to be more creative, more clearly see reality, or put us in contact with other realities ripe for human exploration?

Rick Doblin, who runs the nonprofit Multidisciplinary Association for Psychedelic Studies in Sarasota, Florida, believes that altered states play a role in actually maintaining mental health and that "the brain functions best when it has access to altered states."[47] Richard Glen Boire, director of the Center for Cognitive Liberty and Ethics in Davis, California, believes that intoxication is a basic human right: "Why should it be illegal to alter your style of thinking?"[48] In the Spring of 2005, Nobel-prize-winning economist Milton Friedman, along with five hundred other economists, called for a debate on the current marijuana laws after a Harvard University study found that replacing marijuana prohibition with a system of taxation and regulation, similar to that used for alcoholic beverages, would produce combined savings and tax revenues of between $10 billion and $14 billion a year.

Some may hesitate to change drug laws because of possible health risks, addictions, and car accidents. But what if chemists could design a nonaddictive drug that didn't lead to violence and had minimal health risks, but still permitted us pleasure and insight? Should we still make such a drug illegal?

William James—the psychologist and philosopher whom we discussed in Chapter 3—said, "The world of our present consciousness is only one out of many worlds of consciousness that exist, and that those other worlds must contain experiences which have a meaning for our life." Why do so many people feel certain that our "normal" state is the best one in which to understand the world? Helen Phillips and Graham Lawton in *New Scientist* write:

> If one's view of the world can change so dramatically with the aid of a simple molecule, how can we be sure that our normal brain chemistry is the one most suited to doing science and philosophy? What if our brain chemistry evolved to help us survive at the cost of giving us false beliefs about the world? If so, it is possible that mind-altering drugs might in fact give us a better, not worse, insight than we have in our so-called normal states.[49]

Charles Tart even suggested in a 1972 *Science* magazine article that we create "state-specific sciences," conducted and explored by scientists working and communicating in altered states. He realizes that these kinds of studies have tremendous challenges. For example, people's experiences while in altered states may be ineffable; that is, they may be impossible to communicate them to anyone else. However, many phenomena that are now considered ineffable may be expressible and understandable if we gain more experience, training, and a new vocabulary for communicating transcendent ideas. Tart also recognizes the possibility that various phenomena of altered states of consciousness may be "too complex" for human beings to understand. However, in the history of science, many phenomena that appeared too complex at first were eventually comprehensible.[50]

What is Your Position?

More generally, I'm intrigued by the question of what we should make illegal for all people to use because some people will abuse it. For example, as we discussed, the brilliant scientist, author, and TV personality Carl Sagan said marijuana made him more creative and gave him "devastating insights"—and he thought it was insane and "outrageous" that pot was illegal. Harvard evolutionary theorist Stephen J. Gould also thought it was cruel and misguided to make marijuana illegal to those suffering from cancer and its treatment, as he was. He researched and wrote his critically-acclaimed, 1400-page *The Structure of Evolutionary Theory* while frequently using marijuana to maintain his health.

If America wants to make pot illegal for all because some will abuse it, what penalty should we have given Sagan or Gould for using it? With which drug would you draw the line? Should Carl Sagan have been tossed in jail if he had tried DMT to understand its effect on the mind?

Many of my colleagues have come to believe that marijuana should be legal to those over the age of twenty-one. They suggest that many of the concerns about marijuana smoking are motivated more from moral objections than scientific ones. One friend wrote, "The oft-heard argument that marijuana is a gateway drug was the argument that most made me hesitate. Upon reflection, this argument did not seem any stronger to me than calling beer a gateway to alcoholic beverages." The vast majority of cannabis users do not graduate to more dangerous drugs like cocaine and heroin. Of course, 100 percent of heroin addicts got started on milk, but milk is not a gateway. The dangers of the more potent mind-altering drugs should not be underestimated, but, with the exception of heroin, these drugs contribute to far fewer deaths among their users than either nicotine or alcohol.

The US prisons seem to be needlessly overpopulated with drug offenders. More than five hundred thousand people are housed in federal and state prisons and local jails on drug offenses.[51] The majority of these prisoners are guilty of only minor offenses, such as possessing small amounts of marijuana, or even include people who used marijuana for medicinal purposes. The cost to maintain this great cauldron of imprisoned humanity is more than $10 billion annually.[52] In the early years of the twenty-first century, the federal, state, and local drug-control budgets were almost $40 billion.[53] Nonviolent first offenders caught with only small amounts of a controlled substance frequently are given prison sentences of five to ten years or more.[54] Because prisons are overflowing with such offenders, prisons are forced to release violent prisoners with reduced jail time.[55]

America obviously is not the most repressive nation when it comes to drug laws. People are regularly executed for marijuana offenses around the world.[56] Among the worst offenders are the United Arab Emirates, Saudi Arabia, Burnei, China, and the Philippines. In October 2004, a Malaysian man was sentenced to death in Brunei for possession of a 922 gram slab of cannabis. Under Brunei law, possession of over 600 grams earns the owner the death penalty.[57] Possession of over 500 grams or marijuana usually earns execution in the Philippines. One of my colleagues offers an opinion held by many:

> Recreational drugs should be prescribed by doctors at a reasonable price. Pharmaceutical companies would be responsible for making designer drugs. This would eliminate the dangerous drug trade and make taking drugs safer. The doctors would prescribe the correct dosage and explain the risks. Specialists would have 'safe' rooms where people can be monitored the first time they take a drug, in order to make sure all is well. An interview would follow and the results recorded.[58]

Let's leave this section with a few more facts. The United States has been providing antinarcotics aid to more than dozen countries for more than twenty years, roughly $1 billion a year in recent years.[59] In 1985, the US State Department said that the Peruvian government had destroyed seventy-five hundred acres of coca plants used to make cocaine, but the narcotics trafficking was nonetheless flourishing. In 2005, Peru destroyed twenty-five thousand acres of coca, but coca cultivation continued to rise! Opium poppy production is on the rise in Afghanistan, despite the presence of US troops. The first year after the US overthrew the Taliban, opium poppy production doubled. In 2004, opium production tripled.[60]

Perhaps in an appropriate setting, mind-altering drugs do indeed open doorways that humankind must explore to better understand its place in the universe. Whether these doorways be made available to all adults or to trained professionals, it appears that absolute prohibition walls off scientists, philosophers, and artists to vast vistas of reality. Psychedelic drug researcher Dr. Alexander Shulgin once commented that the amazing phenomena he observed by using mescaline "had been brought about by a fraction of a gram of a white solid, but that in no way whatsoever could it be argued that these memories had been contained within the white solid." He goes on to say, "I understood that our entire universe is contained in the mind and the spirit. We may choose not to find access to it, we may even deny its existence, but it is indeed there inside us, and there are chemicals that can catalyze its availability."[61]

Eternity is a child playing checkers.

—Heraclitus (535–475 BC)

I am surprised to learn that certain police officers have been inclined to minimize the effects of the use of marihuana. [However,] the drug is adhering to its old world traditions of murder, assault, rape, physical demoralization and mental break down. Bureau records prove that its use is associated with insanity and crime. Therefore, from the standpoint of police work, it is a more dangerous drug than heroin or cocaine.

–J. Edgar Hoover, former Director of the Federal Bureau of Investigation[62]

We will pretend that there is nothing hypocritical about two former recreational drug users [Clinton and Bush]–who somehow miraculously emerged unscathed from the billowing clouds of smoke . . . to vie for the most powerful office in the world–who refuse to talk honestly about the issue.

–Gary Kamiya, "Reefer Madness," Salon.com

In 1967, scientific reports from New York University confirmed that banana peels produce no intoxicating chemicals when smoked. Nevertheless, should the government illegalize the smoking of banana peels if the user actually gets high as a result of a psychological placebo effect? For example, scientists have shown that placebos relieve pain by activating the production of endorphins (opioids) in the brain.

–Anonymous

For a movie to make the most money, it has to be seen by the most number of people possible; for it to be seen by many millions of people, it has to connect with something deep down in the collective unconscious. So movies that are huge moneymakers will be the most telling in exploring what is hanging out—and hiding out—in the human psyche.

—Dr. Jane Alexander Stewart, "Movies as Our Cultural Dreams," *Newtopia Magazine*

Despite spending more than $45 billion over the past 25 years and increasing the numbers of people imprisoned for drug offenses more than tenfold, the federal strategy has failed to curb drug use. . . . Drugs are cheaper and easier to get than ever.

—"King's County Bar Association Report"[63]

The mushroom, when eaten, produces an unpredictable range of effects. . . . Psilocybin acts directly upon the language centers of the brain. It facilitates conversation, articulation, and visualization. One wonders what might have impressed a race of human who had not yet developed language. At high doses . . . vocal utterances . . . become visible as fields of changing light patters. Paleolithic shamans would certainly have been impressed by a food which allows vocal vibrations to become visibly beheld. . . . It is possible that not only religion, but language itself sprang forth from the inclusion of *S. cubensis* into the diets of proto-hominids.

—R. Campbell, "You Are What They Ate"[64]

If cannabis was one of the main ingredients of the ancient Christian anointing oil, as history indicates, and receiving this oil is what made Jesus the Christ and his followers Christians, then persecuting those who use cannabis could be considered anti-Christian.

> —Chris Bennet, "Cannabis Linked to Biblical Healing,"
> *BBC News*

In 1475, Turkish law made it legal for a woman to divorce her husband if he failed to provide her with coffee.

> —Helen Phillips and Graham Lawton,
> "The Intoxication Instinct," *New Scientist*

SIX
CLOCKWORK BUTTERFLIES AND ETERNITY
⟡ ⟡

In which we encounter Rainer Maria Rilke, Nathaniel
Hawthorne, beauty, death, robot butterflies, ether butterflies,
hypergraphia, 64 obsession, Nikola Tesla, Philip K. Dick, Alice
Flaherty, writers who used drugs, the dream butterflies of Ernst
Jünger, Jack Kerouac, and Benzedrine-laced Coca-Cola.

*Everything is blooming most recklessly; if it were voices instead of colors,
there would be an unbelievably gorgeous shrieking into the heart of the
night.*

—Rainer Maria Rilke, *Letters of Rainer Maria Rilke,*
1892–1910

The Angels of Rilke

In earlier chapters we discussed my love for the *Epic of Gilgamesh* and
its focus on Gilgamesh's desire to be happy in this world while strug-
gling to learn more about the world to come. Scholars have always
been mesmerized by *Gilgamesh* and its mysterious cast of characters.
For example, my favorite European poet Rainer Maria Rilke
(1875–1926) was in such awe of *Gilgamesh* that he wrote to friends:
"*Gilgamesh* is stupendous! I. . . consider it to be among the greatest
things that can happen to a person. I have immersed myself in it, and
in these truly gigantic fragments, I have experienced measures and
forms that belong with the supreme works that the conjuring Word has
never produced."[1] Rilke described reading *Gilgamesh* "with the fervor
that comes when certain circuits in the brain are firing from an influx
of imaginative power."[2] Even the gaps in the text, where words on the
clay tablets were destroyed by the gentle acid of time, added to Rilke's

sense of "a truly colossal happening and being and fearing." As I immersed myself in *Gilgamesh*, I felt that I was wandering with him through the landscape of dreams or through a mystic afterlife filled with strange beasts and deep forests. I dreamed I was with Gilgamesh at Mount Mashu—smelled the scents of strange roses, heard the cries of lavender birds, and spoke to the scorpion beings that protect the road onward.

Rilke must have also felt the awesome presence of the scorpion beings and stone men. For Rilke, poems opened new realities, and one's imaginary encounters within the poem had lasting effects on the reader's journey in the real world. He felt that to write poetry, one had to have a mystical and close connection with all of life's experiences to make them transcend time:

> Poems are not . . . simply emotions . . . they are experiences. For the sake of a single poem, you must see many cities, many people and things . . . and know the gestures which small flowers make when they open in the morning. You must be able to think back to streets in unknown neighborhoods, to unexpected encounters, and to partings you have long seen coming; to days of childhood whose mystery is still unexplained . . . to mornings by the sea, to the sea itself, to seas, to nights of travel . . . and it is still not enough.[3]

Rilke often thought about Gilgamesh themes—life and its tenuous twisted tether always connecting it to death. Rilke felt that this inevitability of death pushed us "to understand our earthly existence as one side of being, and drain it passionately to the dregs." For Rilke, life was like a flower, blossoming and dying, and perhaps, finally, transforming into a new kind of being, like a caterpillar metamorphosing into a butterfly. No wonder he resonated with Gilgamesh and its ancient avatars.

The leukemia that killed Rilke was cruel. Before he died, ulcerous sores danced in his mouth, his abdomen became painful, and he lost weight. It was during this time that Rilke composed his mysterious epitaph invoking the flower image that transfixed him during his entire life:

> *Rose, oh reiner Widerspruch, Lust, Niemandes Schlaf zu sein unter soviel Lidern.* [Rose, oh pure contradiction, desire to be no one's sleep, under so many lids.]

For Rilke, the rose, through its smell, its beauty, and its transience, symbolized love, death, contradiction, and paradise. Shortly before he composed the epitaph, he had written a poem in French:

> *Est-ce de tous ses pétales que la rose s'eloigne de nous? Veut-elle être rose-seule, rien-que-rose? Sommeil de personne sous tant de paupières?* [Does the rose distance itself from us through all its petals? Does it want to be rose-alone, nothing-but-rose? Sleep of no one beneath so many eyelids?]

Rilke had a bizarre childhood of enforced gender misidentity and abandonment. Like Truman Capote, his mother eventually left him in the care of others. René Karl Wilhelm Johann Joseph Maria Rilke was born prematurely on December 4, 1875. His mother, Sophie, focused on him excessively, perhaps because in the previous year, her daughter died when only a week old. Sophie attempted to resurrect the girl, or at least the concept of having a daughter, by giving Rilke names like René and Maria, which might have sounded slightly feminine or of ambiguous gender to some Europeans. In essence, she turned Rilke into a chameleon. For the first five years of Rilke's life, Sophie called him "Sophia" and dressed him like a girl. Rilke recalled, "I had to wear beautiful long dresses, and until I started school I went about like a little girl. I think my mother played with me as though I were a big doll." Once she could no longer clothe him in dresses, she abandoned him to "the care of a conscienceless, immoral maidservant."

As an adult, Rilke believed in the coexistence of the material and spiritual realms. However, he felt that humans were mere spectators of life, grasping and admiring life's fleeting beauties that were bound to quickly fade away. Yet, he always sought beauty:

> Who, if I cried out, would hear me among the angels' hierar-chies? And even if one of them pressed me suddenly against his heart: I would be consumed in that overwhelming existence. For beauty is nothing but the beginning of terror, which we still are just able to endure, and we are so awed because it serenely dis-dains to annihilate us. Every angel is terrifying.[4]

Nathaniel Hawthorne, Butterflies, 64 Obsession

Rilke was the German language's greatest poet of the twentieth cen-tury. Nathaniel Hawthorne was the preeminent figure in the American literary renaissance of the mid-1800s and one of the greatest masters of American fiction of all time. Rilke and Hawthorne were both inter-ested in the transcendent, ephemeral nature of beauty. For Rilke it was the rose. For Hawthorne, it was the butterfly.

Hawthorne was born in 1804 in Salem, Massachusetts. As with Truman Capote and other famous authors, Hawthorne decided at an early age that he would become a writer. After graduating from Bow-doin College, he wrote several successful short stories. My personal favorite is "The Artist of the Beautiful" (1844), a little-known and haunting tale of a man who creates a lifelike, mechanical butterfly for the woman he loves—a story that I'll discuss later in this chapter. Alas, Hawthorne couldn't support himself with the little money he made from selling short stories, so he took a job as surveyor of the Port of Salem, but was fired after a few years. "I detest this town so much," Hawthorne said, "that I hate to go out into the streets, or to have people see me."

Hawthorne had an unhappy start to life. His father, a ship captain, died at sea when Hawthorne was four, and ever after, Hawthorne seemed to have a gloomy view of life. In his novel *The House of the Seven Gables*, he writes, "Not to be deficient in this particular, the author has provided himself with a moral—the truth, namely, that the wrongdoing of one generation lives into the successive ones." Perhaps Hawthorne was feeling guilty that he was a descendent of John Hawthorne, one of the judges in the Salem witchcraft trials of 1692.

Hawthorne was frequently unhappy with his early works. For example, he burned his first short-story collection, *Seven Tales of My Native Land*, after publishers rejected it. He burned unsold copies of his first novel *Fanshawe*, which was based on his college life.

In 1842, Hawthorne married Sophia Peabody, who, like Hawthorne was rather reclusive. Sophia fell in love with his work immediately, and she wrote in her journal, "I am always so dazzled and bewildered with the richness, the depth, the . . . jewels of beauty in his productions that I am always looking forward to a second reading where I can ponder and muse and fully take in the miraculous wealth of thoughts."[5] Hawthorne and Sophia had three children: Una, Julian, and Rose. Alas, Una was mentally ill and died young. Interestingly, many eminent persons have had children with serious mental problems or had at least one child take his or her life. One of Robert Frost's daughters was committed to the state mental hospital and another daughter had "nervous breakdowns." One of Albert Einstein's children was diagnosed as schizophrenic. Ambrose Bierce's oldest son committed suicide and his other died of alcoholism at age twenty-seven. Thomas Edison had two children who became alcoholics, one of whom committed suicide. Alfred Stieglitz's daughter was psychotic and committed to a mental institution. James Joyce had two children. His son became an alcoholic; his daughter went mad and was admitted to an asylum for schizophrenia.

Despite the unfortunate setbacks in Hawthorne's early life, his *The Scarlet Letter*, published in 1850, was finally a resounding success. The story focuses on the illicit love affair of Hester Prynne with the Reverend

Arthur Dimmesdale, with whom she has a baby. Prynne is forced to wear the scarlet "A" because she has committed adultery. The subject was so popular that the first edition sold out within ten days of publication. Additional editions came out in quick succession, and the work became popular reading throughout the United States.

The Scarlet Letter continues to be of interest. Demi Moore starred in the 1995 movie version, and in 2005, an unnamed American bookseller purchased the book's original printing proofs for $545,000! Hawthorne had burned the original manuscript after the publisher returned it to him. The proofs contain handwritten comments by both the editor and Hawthorne.[6]

Toward the end of his life, Nathaniel Hawthorne compulsively wrote the number "64" on any scrap of paper he saw. Scholars have ever since wondered what secret meaning, if any, this might have. Some people say he accurately predicted the year of his death, 1864, when he died while on a trip to the mountains with his presidential friend Franklin Pierce. Perhaps he was thinking of the 64 squares on a chessboard. Others say this behavior was simply the result of obsessive-compulsive disorder, with no hidden meaning at all.

The Artist of the Beautiful

One particularly mystical Hawthorne story is "The Artist of the Beautiful"—little known except to Hawthorne scholars. The story, published in 1846, centers on the life of genius Owen Warland who works in a watch shop. Owen is a sensitive young man, who is secretly in love with Annie Hovenden, the shopkeeper's daughter. Owen spends his time wondering if it is possible to create life in the form of a beautiful butterfly. Hawthorne describes Owen:

> He was attempting to imitate the beautiful movements of Nature, as exemplified in the flight of birds or the activity of little animals. It seemed, in fact, a new development of the love of the

Beautiful, such as might have made him a poet, a painter, or a sculptor. . . . The character of Owen's mind was microscopic, and tended naturally to the minute, in accordance with his diminutive frame, and the marvelous smallness and delicate power of his fingers.[7]

Owen eventually succeeds in making the butterfly. Perhaps the flying creature is powered by springs and circulating fluids, and makes use of tiny watch-like components—the reader is never told precisely how it operates. The shopkeeper discovers an early model and almost accidentally crushes it:

"But what is this?" cried Peter Hovenden abruptly, taking up a dusty bell-glass, beneath which appeared a mechanical some-thing, as delicate and minute as the system of a butterfly's anatomy. "What have we here! Owen, Owen! There is witch-craft in these little chains, and wheels, and paddles! See! with one pinch of my finger and thumb, I am going to deliver you from all future peril."

"For Heaven's sake," screamed Owen Warland, springing up with wonderful energy, "as you would not drive me mad, do not touch it! The slightest pressure of your finger would ruin me forever."[8]

The butterfly is saved, but, sadly, Annie eventually marries someone else. In the closing scene of the story, Owen decides to show her the new life he has created.

Owen produced . . . a jewel-box. It was carved richly out of ebony by his own hand, and inlaid with a fanciful tracery of pearl, representing a boy in pursuit of a butterfly, which, elsewhere, had become a winged spirit, and was flying heavenward; while the boy, or youth, had found such efficacy in his strong desire, that he ascended from earth to cloud, and from cloud to celestial atmosphere, to win the Beautiful.

. . . A butterfly fluttered forth, and, alighting on her finger's tip, sat waving the ample magnificence of its purple and gold-speckled wings, as if in prelude to a flight. It is impossible to express by words the glory, the splendor, the delicate gorgeousness, which were softened into the beauty of this object. Nature's ideal butterfly was here realized in all its perfection; not in the pattern of such faded insects as flit among earthly flowers, but of those which hover across the meads of Paradise, for child-angels and the spirits of departed infants to disport themselves with.

Owen and Annie contemplate the exquisite detail of the mechanical butterfly, and Annie wonders if the creature is truly alive or a mere clockwork mechanism.

The rich down was visible upon its wings; the luster of its eyes seemed instinct with spirit. The firelight glimmered around this wonder—the candles gleamed upon it—but it glistened apparently by its own radiance, and illuminated the finger and outstretched hand on which it rested, with a white gleam like that of precious stones. In its perfect beauty, the consideration of size was entirely lost. Had its wings overreached the firmament, the mind could not have been more filled or satisfied.

"Beautiful! Beautiful!" exclaimed Annie. "Is it alive? Is it alive?"

The butterfly now flung itself upon the air, fluttered round Annie's head, and soared into a distant region of the parlor, still making itself perceptible to sight by the starry gleam in which the motion of its wings enveloped it. The infant, on the floor, followed its course with his sagacious little eyes. After flying about the room, it returned, in a spiral curve, and settled again on Annie's finger.[9]

In the end of the story, an uncaring child crushes the butterfly in his hand. Annie screams as she looks into the child's hand at a "small heap of glittering fragments." As Owen looks at the ruin of his life's labor, he has an epiphany of sorts. The destruction of his lovely creation, however distressing to Annie, does not seem to upset Owen. Although the butterfly was dead, the transcendent, timeless reality of the butterfly and its beauty persisted beyond the extinction of its mechanical form:

> Owen had caught a far other butterfly than this. When the artist rose high enough to achieve the Beautiful, the symbol by which he made it perceptible to mortal senses became of little value in his eyes, while his spirit possessed itself in the enjoyment of the Reality.[10]

Let us return to Hawthorne's repeated and mysterious writing of "64." This kind of compulsion is not as odd as it may seem among geniuses. For example, the electrical genius Nikola Tesla had "arithromania" or "numerical obsessive-compulsive disorder." He demanded precisely eighteen clean towels each day. If asked why, Tesla provided no explanation. Table accoutrements and towels were not the only items he demanded come in multiples of three. For example, he often felt compelled to walk around the block three times. He always counted his steps while walking. He chose room number 207 in the Alta Vista Hotel, because 207 is divisible by 3.

In *Strange Brains and Genius*, I catalogue numerous examples of numerical obsession, most of which come from Judith Rapoport's *The Boy Who Couldn't Stop Washing*. For example, an eighteen-year-old boy is compelled to count to twenty-two over and over again. He taps on the wall twenty-two times or in multiples of twenty-two. He walks through doorways twenty-two times and gets in and out of his chair twenty-two times. The boy becomes addicted to drugs that have interesting effects on his twenty-two -ness. For example, while on amphetamines and cocaine, his twenty-two tapping increases to the point where all his time is spent tapping out twenty-two all over his walls. LSD makes the ritual completely disappear. For a more complete list of kids with numerical obsessive-compulsive disorder, see *Strange Brains and Genius*.[11]

Hawthorne's repeated writing of 64 is perhaps a specific and counterproductive form of hypergraphia—obsessive writing that can be a sign of certain forms of epilepsy. Although hypergraphia is often nonproductive, it has been used constructively by temporal lobe epileptics. For example, best-selling science-fiction writer Philip K. Dick—who wrote the stories on which the movies *Total Recall, Blade Runner, Impostor, Barjo, Paycheck, Screamers*, and *Minority Report* are based—probably had epilepsy and hypergraphia. From age fifteen, Dick suffered from auditory and visual hallucinations that he interpreted as signs from God. He also had macropsia and micropsia (where stationary objects like chairs appear to enlarge or shrink), and depersonalization. Many of Dick's symptoms found their way into his books in the characters who had hallucinatory experiences. Dick published his first short story at age thirteen writing at incredible speeds, and later produced thousands of pages of handwritten journals in addition to his forty-one books.

Another manifestation of obsession and hypergraphia is found in compulsive diary writers. Perhaps the most famous case of very detailed diaries was reported in a *Seattle Times* feature that chronicled Robert Shields, seventy-seven, of Dayton, Washington. Shields was the author of a thirty-eight-million-word personal diary. Filling eighty-one cardboard boxes, the diary allegedly covered the past twenty-four years of his life in five-minute increments. An example: *July 25, 1993, 7 A.M.* , "I cleaned out the tub and scraped my feet with my fingernails to remove layers of dead skin." *7:05 A.M.* , "Passed a large, firm stool and a pint of urine. Used five sheets of papers."[12]

Geniuses and normal people have often had secret obsessions and compulsions. A. J. Jacobs, author of *The Know-It-All*, admits that he has a secret radio ritual. Whenever he turns off the radio, the last word he hears must be a noun. He writes,

> No verbs, no prepositions, no adjectives—I need a noun, prefer-
> ably a good, solid noun, something you can hold in your hands.
> So I'll stand over my shower radio, dripping, pushing the power
> button on and off and on and off till I catch Nina Totenberg
> saying something like "bottle" or "car." Only then can I get out
> of the shower and get dressed.[13]

Another famous example of constructive hypergraphia is the case of Harvard Medical School neurologist Alice Flaherty. Following the deaths of her twin boys, she awoke one day with an overwhelming desire to write down all her thoughts. The urge was so powerful that she took medications to suppress the writing, but these drugs were not very effective. To help explain her affliction to a wide audience, she has since published several books including *The Midnight Disease: The Drive to Write, Writer's Block, and the Creative Brain*. She has come to believe that for her, writing and writer's block arise from interactions between spe-cific areas of the brain. The brain's limbic system provides the emo-tional push to write. Many nerve fibers connect it to the temporal lobes, brain areas near the ears that interpret words and give rise to ideas. The frontal lobe, behind the forehead, is the efficient organizer and editor.[14] If the temporal lobe becomes hyperactive, a person may rapidly produce hundreds of wordy sentences.

Someday we may be able to harness the positive powers of hyper-graphia through electrical stimulation of the brain. Moreover, Fla-herty describes one of her patients who periodically administers electric currents to the brain in the hopes of ameliorating the tics pro-duced by Tourette's syndrome. At certain current settings, the woman reported increased creativity and productivity in both her professional and personal activities. "Her boss asked her what they had put in her head, and how he could get one."[15] For a few creative people, drugs have kick-started beneficial hypergraphia. Robert Louis Stephenson wrote *The Strange Case of Dr Jekyll and Mr Hyde* during a six-day cocaine-binge.

We spoke of the numerous, famous, alcoholic writers in Chapter 1. Narcotics, stimulants, and hallucinogens have also inspired and influenced writers through the ages. As just a few examples, consider: opium (Charles Baudelaire, Thomas De Quincey, Oscar Wilde, John Keats, Jean Cocteau, William S. Burroughs, Elizabeth Barrett Browning, Percy Shelley, Samuel Taylor Coleridge, Edgar Allan Poe), cannabis and hashish (Charles Baudelaire, Arthur Rimbaud, Gustave Flaubert), cocaine (Robert Louis Stephenson, Sigmund Freud, Sir Arthur Conan Doyle), speed and LSD (Jack Kerouac, William S. Burroughs, Timothy Leary, Michel Foucault, Anaïs Nin), and mescaline/peyote (Henri Michaux, Antonin Artaud, Allen Ginsberg).[16]

William Burroughs, author of *Naked Lunch*, once said that he never regretted his experience with drugs: "I think I am in better health now as a result of using junk at intervals than I would be if I had never been an addict. When you stop growing you start dying. An addict never stops growing." No doubt Burroughs's drug use helped shape his perceptions and contributed to some his famous stream of conscious musings in which human beings transform into pulsating blobs of protoplasm, covered with larvae:

> Doolie, sick, was an unnerving sight. The envelope of personality was gone, dissolved by his junk-hungry cells. Viscera and cells, galvanized into a loathsome insect-like activity, seemed on the point of breaking through the surface. His face was blurred, unrecognizable, at the same time shrunken and tumescent.[17]

In the 1950s, Burroughs sought the psychedelic brew known as yagé or ayahuasca in South America, and he later wrote, "There is nothing to fear. Your ayahuasca consciousness is more valid than normal consciousness. . . ." While in this altered state, he saw "larval beings" that passed before his eyes "in a blue haze."

Poet, essayist, playwright, actor, and director Antonin Artaud sometimes credited his creativity to the *absence* of opium: "It is not opium which makes me work but its absence, and in order for me to feel its absence it must from time to time be present."

Many readers may be aware of my fondness for French novelist Marcel Proust (1871–1922) and his monumental book *In Search of Lost Time.* Some scholars suggest that his visions and stream-of-consciousness techniques were drug-inspired. After all, in *The Guermantes Way,* Proust describes his "secret garden in which the different kinds of sleep, so different from one another, induced by datura, by Indian hemp, by the multiple extracts of ether—the sleep of belladonna, of opium, of valerian—grow like unknown flowers whose petals remain closed until the day when the predestined stranger comes to open them with a touch and to liberate for long hours the aroma of their peculiar dreams for the delectation of an amazed and spellbound being." Marcus Boon writes in *The Road to Excess,*

> Proust's mind and body were constantly awash in a sea of chemicals that produced precisely the kind of cognitive movements that he describes in his books. . . . The extreme form of literary transcendence that Proust, sitting in his cork-lined room for years, exploring his own interiority, was a mascot for, could not be sustained without some level of chemical support. Nor was it.[18]

Another French writer, René Daumal, was a butterfly collector who happened to use carbon tetrachloride to kill and preserve his specimens. He also sniffed the fumes to reveal a higher reality:

And this "world" appeared in its unreality, because I had abruptly entered another world, intensely more real, an instantaneous world, eternal, a fiery inferno of reality and evidence into which I was thrown, spinning like a butterfly in a flame. At that moment, there is *certainty*, and it's here that words must be content to circle around the bare fact. [19]

In his altered state, Daumal felt that sounds were somehow required to uphold the fabric of reality. Daumal had to repeat certain sounds to solidify his presence in these realms, and sounds play a role in other people's DMT visions as well. He writes:

A sound accompanied this luminous movement, and I suddenly realized it was I who was making it. In fact, I virtually *was* that sound; I sustained my existence by emitting it. The sound consisted of a chant or formula, which I had to repeat faster and faster. . . . The formula . . . ran something like this:

Tem gwef rem gwef dr rr rr.[20]

The Polish writer and painter Stanislaw Ignacy Witkiewicz experimented with drugs like peyote and cocaine and actually annotated each of his paintings with the name of the drugs he was taking while executing the painting. Witkiewicz exhibited his genius in early childhood—reading scientific and philosophical works in various languages and writing short comedies in imitation of Shakespeare at the age of seven. Through his life, he developed various quirky behaviors such as obsessively washing his hands. When Soviet troops invaded Poland, he killed himself in protest against the dreaded regime. Years later, when the Ministry of Culture of Communist Poland decided to exhume his body, to give it a VIP burial, they found a woman's corpse in the coffin, apparently part of a joke that Witkiewicz intended.

Benzedrine and Coca-Cola Dreams

Narcotics, stimulants, and hallucinogens have shaped some of humanity's most fundamental philosophies and have provided a significant portion of its economic wealth. Until the 1900s, there was little shame attached to using drugs to help adjust to life's stresses and problems or to foster productivity and creativity at work. French novelist Charles Nodier said that opium gave him revelations that he could not obtain when sober. Flaubert, one of many Paris hashish users of the 1800s, said, "We were troubadours, rebels—above all, we were artists." French writer Honoré de Balzac claimed to have "heard celestial voices and seen heavenly paintings" after sampling hashish. Poet W. H. Auden took Benzedrine, an amphetamine, every morning as a "labor-saving device" in his "mental kitchen." Graham Greene was also a heavy user of Benzedrine.

Of course, stimulants can have deleterious effects. Richard Davenport-Hines in "Do Artists Need Narcotics Even More than Ordinary People?" comments on the downside of drug use: "Jean-Paul Sartre, who used amphetamines while writing a 50-page essay on Jean Genet, and ended up with an 800-page book, and whose 3,000-page study of Flaubert is definitively unreadable, demonstrated the dangers of writing on speed."[21]

Jack Kerouac used to dump the contents of Benzedrine nasal inhalers into Coca-Cola and drink the laced brew. According to legend, while using Benzedrine and coffee, Kerouac wrote the novel *On the Road* in only two weeks in a long session of spontaneous prose. For this book and others, he fed a long roll of paper into his typewriter and just kept typing until the manuscript was around one hundred feet long. In 1953, he completed *The Subterraneans* in three nights using Benzedrine and a long typewriter roll. He felt Benzedrine "intensified his consciousness" but realized that overuse of the drug was dangerous: "I'm taking enormous doses of Benzedrine to write my novels—I probably won't live long enough to enjoy my money."[22] He died at age forty-seven.

The Dream Butterflies of Ernst Jünger

Let us conclude this chapter with a quotation that marvelously intertwines several pervasive threads of this book—butterflies, drugs, brains, and dreams. The words are from the prolific writer Ernst Jünger (1895–1998), the son of a wealthy German chemist and author of *Heliopolis: A Novel* and *Approaches to Drugs and Intoxication*, in which he describes his own drug experiments.

Jünger was the consummate chameleon, having served as a lieutenant in the German army while also devoting himself to the study of marine biology, botany, and philosophy. He attained international prestige as a beetle-expert with his personal collection of over forty thousand specimens! Throughout his very long life, he experimented with a variety of drugs, including LSD, mescaline, ether, cocaine, and hashish.

Take a deep, slow breath, and enjoy an excerpt from Ernst Jünger's *Heliopolis*:

> He captured dreams, just like others appear to chase after butterflies with nets. He did not travel to the islands on Sundays and holidays and did not frequent the taverns on Pagos beach. He locked himself up in his studio for trips into the dreamy regions. . . . The drugs served him as keys to entry into the chambers and caves of this world. He went on voyages of discovery in the universe of his brain.[23]

I reproach all modern religions for having presented to their faithful the consolations and extenuations of death, instead of giving their souls the means of getting along with death and coming to an understanding of death, with its complete and unmasked cruelty.

—Rainer Maria Rilke, *Selected Letters of Rainer Maria Rilke*

Eckels felt himself fall into a chair. He fumbled crazily at the thick slime on his boots. . . . "No, it can't be. Not a little thing like that. No!"

Embedded in the mud, glistening green and gold and black, was a butterfly, very beautiful and very dead. . . . It fell to the floor, an exquisite thing, a small thing that could upset balances and knock down a line of small dominoes and then big dominoes and then gigantic dominoes, all down the years across time . . . Killing one butterfly couldn't be that important! Could it?

—Ray Bradbury, "A Sound of Thunder"

I talk to a lot of people who've never had psychedelic experiences . . . or who have never even smoked pot—and they still seem just as aware of the fact that we're all living in reality tunnels, and that we choose different tunnels.

—Douglas Rushkoff, in David Jay Brown's *Conversations at the Edge of the Apocalypse*

I could see a new world with my middle eye, a world I had missed before. I caught images behind images, the walls behind the sky, the sky behind the infinite.

—Anaïs Nin, *Diaries*

The hallucinogen drugs shift the scanning pattern of "reality" so that we see a different "reality." There is no true or real reality. Reality is simply a more or less constant scanning pattern.

–William Burroughs, *Nova Express*

Snakes too good to be true. Pearly tanks of the Assyrian Kings. Cross-section of reptile machine.

–Polish writer and painter, Stanislaw Ignacy Witkiewicz, while writing under the influence of peyote, 1928

Who is Ernst Jünger? The body of Jünger's work contains more than fifty volumes. His novels, stories, journals, and collections of aphorisms have won countless literary honors and prizes. Unfortunately he is little known in the United States. But in a long and adventurous life, Jünger has been able to fulfill two-thirds of the famous prescription of Baudelaire: "there are but three beings worthy of respect: the priest, the warrior, and the poet. To know, to kill, and to create."

–B. John Zavrel, "Ernst Jünger: A Miracle at 100 Years"[24]

. . . SETI [the search for extraterrestrial intelligence] is looking in the wrong direction. If . . . we see an anthill, do we go down to the ant and say, "I bring you trinkets, . . . I bring you knowledge, . . . I bring you nuclear technology, take me to your leader"? Or, do we simply step on them? . . . Would the ants even know what a ten-lane superhighway is . . . or how to communicate with the workers who are just feet away?

–Michio Kaku, "Parallel Universes, the Matrix, and Superintelligence," KurzweilAI.net[25]

When Coleridge's "Kubla Khan," Shelley's *Frankenstein*, and Robert Lewis Stevenson's *Strange Case of Dr. Jekyll and Mr. Hyde* were published as transcripts of dreams, did this make their writers into authors or transcribers, perhaps even fraudsters or plagiarists, misrepresented as the authors of their work? All writers on drugs become ghostwriters for their drugs. Or perhaps their drugs are ghostwriting them.

–Sadie Plant, *Writing on Drugs*

Seven

Evolution, Ice Cream, and the Goddess of Chopped Liver

✣ ✣

In which we encounter Sylvia Weinberger, chopped-liver, God's laryngeal nerve, Reuben Mattus, "prochronic" events, Häagen-Dazs ice cream, the Reese Candy Company, livers in myth, foie gras, Liver-Eating Johnston, *Silence of the Lambs, Attack of the Liver Eaters*, "Merrye Syndrome," liver evolution, liver divination, intelligent design, creationism, molecules in space, the emergence of life and new species, monkeys typing the Bible, Robert Ardrey's killer ape, mosquitoes, flowers, polyploidy, beetle engineering, poodles and wolves, fecund enclaves of subterranean creatures, the Omphalosian view of reality, Philip Henry Gosse, John C. Whitcomb, Arkansas Act 590, abortion, escape ovulation, zygotes, hepatoscopy, Etruscans, haruspimancy, the liver and the "butterfly effect," divination, and Shakespeare.

The liver, that great maroon snail: No wave of emotion sweeps it. Neither music nor mathematics gives it pause in its appointed tasks.
—Richard Selzer, *Mortal Lessons: Notes on the Art of Surgery*

The Chopped-Liver Queen

Very few of us thrive and become wealthy by following our bliss, or by turning a small part of our lives into something grand. Sylvia Weinberger (1906–1995) transcended both time and space by becoming the chopped-liver queen. The seeds of her success were planted in 1944 when she started making chopped liver for her Bronx, New York, luncheonette.[1] Within a few years, she was the head of a $2-million-a-year company called "Mrs. Weinberg's Chopped Liver." Her liver—made from broiled beef liver, onions sautéed in oil, matzo meal, salt, pepper, and hard-boiled eggs—was the leading brand of chopped liver at kosher restaurants and homes by the time she retired in 1989.

Her business method was simple and direct. First, she coaxed her luncheonette customers to take the chopped liver home. They soon found that they loved the liver and became virtually addicted to its taste. They scrambled for more. Within months, demand for the product exploded in Bronx supermarkets that Sylvia had solicited. She had no special containers for the liver at this point in her career and had to make do with whatever was available at home or in her store. Gargantuan orders for the chopped liver came to her so rapidly that she had to stuff the adipocere-like amalgam into large plastic bags for delivery.

Weinberger represents the classic rags-to-riches American story of someone who had an idea, persevered, and soared to stardom. She had come to the United States from Hungary in 1937, worked long hours, bought the luncheonette, and finally sold it in 1955 to create chopped liver as a full-time, all-consuming operation. Weinberger's chopped liver is one of many Bronx food products that took off like a rocket in the mid-1900s after being spearheaded by a single individual.

Another Bronx success story is that of Häagen-Dazs ice cream. Contrary to popular thinking, Häagen-Dazs was not created or named in a Scandinavian country. Rather, Polish-born Reuben Mattus of the Bronx invented this European-sounding name in 1961—to give his own brand of ice cream an international, Old-World association. "Häagen-Dazs" has no particular meaning, but it sure does sell. Mattus's interest in ice cream started in the 1920s, when he worked for his mother to help her make money from her homemade ice cream. They traveled routes around the Bronx, selling the ice cream from their horse-drawn wagon.

According to Rose Mattus's blockbuster book, *The Emperor of Ice Cream: The True Story of Häagen-Dazs, A Love Story*, the thick creamy Häagen Dazs resulted from a factory mishap, when the air injection pump broke. The accidental product tasted superior to any ice cream that was on the market at the time. When formulating the ice cream name, Reuben Mattus had said, "I think maybe a Danish name. . . . They're nice people you know. Good people. They tried hard to save Jews during the war, ferried them to safety ahead of the Nazis. . . . Everyone likes the Danes." The name was also a Danish-sounding inversion of the popular Duncan Hines.

When you consider the recent history of ice cream, the Jews reign supreme. Not only is there Rose and Reuben Mattus of Häagen-Dazs fame, but we also have Ben Cohen and Jerry Greenfield of Ben and Jerry's. They're famous for inserting unusual objects in their ice cream, like crushed toffee bars and pretzels, which made for a more daring product. Before his ice cream stardom, Jerry was known for his biological interests and experiments to analyze oxidative phosphorylation in beef heart mitochondria. Ben worked as a pediatric emergency room clerk on the night shift at Bellevue Hospital before hitting it big with ice cream.

Other gastronomical royalty include the makers of various candies. In 1923, the Reese Candy Company (famous for Reese's Peanut Butter Cup) started in the basement of Harry Reese's home. He worked as a farmer during the day to support his wife and sixteen children! At night, he produced the candy. Today the Reese's Peanut Butter Cup is a phenomenal seller. Harry Reese died in 1956, and a few years later, the H. B. Reese Candy Company, Inc., was sold for $23.5 million to the Hershey Chocolate Company, currently known as Hershey Foods Corporation.

Hans Riegel invented gummi bears (the first gummi candy) during the 1920s. Forrest Mars, Sr., invented the recipe for M&M candies during the Spanish Civil War, but M&Ms were not sold to the public until 1941, when they were packaged in cardboard tubes. Cheerios, America's first ready-to-eat oat cereal with doughnut-like rings also appeared on grocery shelves in 1941.

Chopped liver never gained the popularity of these candies or cereals, but Jews continue to eat chopped liver with great gusto, particularly during the holidays and with family. According to chopped-liver expert Andy Lax, chopped liver "is a cultural icon of both industriousness and indulgence. Spread thickly on a slab of chewy rye or layered with corned beef and coleslaw, it transforms the plain into the sublime."[2] If we wish to avoid the high fat content of chopped liver, some health nuts create vegetarian "chopped liver" where mushrooms or lentils replace the liver.

Today, liver has made its way into the most unlikely products. For example, Oscillococcinum, a homeopathic remedy for flu symptoms available in drug stores, is made from super-diluted extract of duck liver—and $20 million worth of Oscillococcinum was sold in 1996 alone! The liver is so dilute in each pill, that a single duck could be used to make a world supply for one year. Oscillococcinum is the best-selling over-the-counter flu medication in France.

Liver and Legend

Humans probably ate animal livers since the dawn of humanity. One of the earliest tales of liver eating goes back to ancient Egypt in the famous myth of two brothers, one named after Anubis, the jackal-headed god of the afterlife, and the other named after Bata, who often took the form of a bull. Anubis's wife, displeased with Bata, asks Pharaoh to let her eat his liver when Bata is in his bull manifestation.

In ancient Greek mythology, Zeus chained Prometheus to a rock and made him endure the torture of an eagle penetrating his body and eating his liver every day. Prometheus's liver would promptly grow back every night.

Here's another liver factoid for you. Most Americans have heard of the phrase, "What am I—chopped liver?" which was coined in America. Because chopped liver is a side dish and never a main course, the phrase is used to express hurt when someone feels overlooked, that is, treated as a "side dish."

Authentic foie gras is made from goose liver, and the technique of optimizing its taste is over four thousand years old, from the fourth and fifth Egyptian dynasties. Archeologists have found paintings in

Egyptian tombs that depict geese being held by the neck and being fed balls of grain. This odd technique, called gavage, relies on the fact that overfed or force-fed geese eventually produce oversized livers that are tastier than normal.

Liver-Eating Johnston

Every time I sit down to eat a bit of ordinary chopped liver with matzos, a little shiver goes up my spine for two reasons:

1. I like the taste.
2. I think of John "Liver-Eating" Johnston

John "Liver-Eating" Johnston's date of birth is unknown. He died on January 21, 1900. The imposing figure of a man stood six feet six inches tall and weighed over 240 pounds. For most of his life, he lived in the Wyoming mountains, making a living in fur trapping and selling chopped wood to steamboats.

Johnston was famous for his hatred and love of Native Americans. For example, when a woman and her family living in the Musselshell River basin of the Rockies were slaughtered by local Indians, Johnston tracked down and killed the assailants. In 1847, Johnston's pregnant Indian wife was murdered and scalped by a raiding party of Crow Indians while he was away hunting. Johnston went wild, fought the Crow Indians, and ate the raw livers of his slain enemies to finalize his conquest. This gross practice earned him the title of "Liver-Eating" Johnston. For more than twenty years, he maintained a solitary battle with the Crows. Years later, after his thirst for revenge and justice had abated, one of his friends was murdered by the Crows, and Johnston killed all of the Indians involved and the traders who sold them their rifles. He was feared and respected in old age, and served in various law enforcement capacities. Today, you can visit his grave in Trail Town, a tourist "pioneer village" in Cody Park County, Wyoming.

These days, a controversy rages as to whether Johnston ate as many livers as legends suggest. At times, Johnston even denied ever being married to an Indian woman or eating the liver of a single human. However, when Johnston made his denials, he was a sheriff and perhaps was trying to paint a more pristine image of himself. Most of the life of Johnston, as recounted in contemporary books, is derived from oral tradition and fragments of writing, almost like the *Epic of Gilgamesh*. Most scholars today do not doubt that this big man had a long-running feud with the local Indians. The exact number of livers he ate will forever remain a mystery.[3]

Liver eating has been featured in a number of movies, most famously in the predatory character of Hannibal Lechter who appears in three Thomas Harris novels, *Red Dragon*, *The Silence of the Lambs*, and *Hannibal*. In the 1991 film version of *Silence of the Lambs*, the murderous Lechter speaks his famous line, "I ate his liver with some fava beans and a nice chianti."

The creepiest movie ever made with "liver" in its title was the 1964 black-and-white movie *The Liver Eaters*—also known as *Spider Baby* or *Attack of the Liver Eaters*—which features the Merrye siblings who are afflicted with "Merrye Syndrome," a neurological disorder that begins to manifest itself at the age of ten, causing the brain to slowly decay until the brain is no more developed than a fetus's brain. This cult movie features Ralph, who eats "anything he can catch," and Virginia who imagines herself to be a human spider. She catches her prey in a weblike net and then bites them with her "fangs" (two long knives). Bruno, played by Lon Chaney, Jr., is a kindly man who sympathizes with the children's affliction and seems to tolerate them to a degree. *Attack of the Liver Eaters*, directed by former Roger Corman production assistant Jack Hill, was a low budget project and filmed in only twelve frantic days. It didn't do well in its opening year, but years later, it played at double features in drive-in movie theaters. Once it appeared on home video in the early '80s, the movie developed a cult following.

What is the Liver?

Walk up to the average adult on the street and ask him or her, "What is the function of the liver?" Most would not be able to answer. I've tried this experiment and found that 80 percent did not know the several functions of the liver. Try the question on friends. The liver is indeed a complete mystery to most people.

The liver exists in all vertebrates and detoxifies the blood and produces bile for digestion. It performs a number of other nifty functions as well—for example, it stores glycogen (a carbohydrate) and synthesizes blood plasma proteins. Nature protects the organ with a thin, double-layered membrane that reduces friction against other organs. Scientists cannot currently engineer an artificial liver, so, alas, you can't live without it.

Sometimes, as my mind drifts off to sleep late at night, I wonder when the liver evolved. I know, for example, that arthropods have a digestive gland that functions a little like a liver, and in insects this organ is known as the "fat body." Cephalochordates (small aquatic animals) are the closest invertebrate relatives to vertebrates—and these creatures have a caecum, a digestive organ that resembles both the liver and pancreas in vertebrates.

Liver evolution, and evolution in general, fascinates me. Let's try to decide in the next few pages if evolution is fact or fiction.

Evolution: Fact or Fiction?

In March 2001, the Gallup News Services reported that 45 percent of Americans agree with the statement, "God created human beings pretty much in their present form at one time within the last 10,000 years or so."[4] In a forced choice between creationism and evolution to explain the diversity of life on Earth, 57 percent chose creationism. However, evolution is a *fact* if we take the word to mean that living organisms change through the process of replication, variation, and selection, and that this change takes place over long periods of time. Fossil evidence and biological experiments suggest that organisms adapt to new ecological niches and changing conditions through genetic mutation. For those of you who prefer the word "theory" to "fact" when discussing evolution, please remember that when scientists use the word "theory," they refer to a logical, tested explanation for a great variety of facts (e.g., "the theory of relativity"), and the term "theory" does not mean an unsubstantiated guess or assumption. In any case, evolution appears to have more explanatory power than any other competing mechanisms for how life on earth changed through billions of years.

Creationists who reject evolution often argue that sufficient fossil evidence does not exist that links one animal form to the next. What they don't seem to realize is that evolution is suggested by many different kinds of evidence, and it is not surprising that we can't find fossil links between every two animal forms a creationist should suggest. Michael Shermer, *Scientific American* columnist, has called this continual pointing to gaps in the fossil record "the fossil fallacy of creationists." To clarify his point, he notes that with so many breeds of dogs popular for so many thousands of years, one would think there would be an abundance of transitional skeletons providing scientists with reams of data from which to reconstruct the dog's evolutionary development.[5] However, the evidence based on bones is sparse. On the other hand, studies of mitochondrial DNA from early dog remains suggest that dogs have a common origin fifteen thousand years ago in gray wolves from East Asia. Shermer's point is that we know evolution happened because of a convergence of various kinds of evidence that includes findings in molecular biology, geology, paleontology, and comparative anatomy. Evolutionary biologist Richard Dawkins likens the scientific process to a detective visiting a crime scene and working out what must have happened by examining the clues. Evolution has countless clues including the distribution of fossils and DNA codes in organisms.

Of course, biologists are aware of numerous transitional species that creationists may once have referred to as missing links. Consider as just one example *Tiktaalik*, the creature discovered in 2006 that represents an intermediate between finned fish and four-footed animals called terapods. Its fins actually contain the beginnings of a terapod hand, complete with a simple version of a wrist, elbow, and five fingerlike bones. It has a flexible neck and eyes mounted on top of its head—like a crocodile—and a large ribcage that suggests that *Tiktaalik* had lungs that would permit it to live on land for short durations.

Why don't we find even more fossil evidence of missing links? Perhaps transitional species, created by the adaptive genetic changes necessary to evolve a new dominant species, will be rare. As an analogy, imagine a future archeologist looking for the transition between the bicycle and the motorcycle in a large garbage dump where two-wheeled vehicles have been tossed for over a century. On the bottom layer, he would find many bicycles, and above, he would find many motorcycles. But there is very little chance that he would discover one of the first experimental breeds of motorcycles such as the two-cylinder, steam-engine motorcycle (powered by coal) in 1867—or the first gas-engined motorcycles in 1885, which were engines attached to wooden bikes.

Although many famous Christians have believed in evolution, including Pope John Paul II who said evolution has been "proven true," some of my creationist friends reject evolution by asserting that complex structures in cells, or complex biomolecules, could not come together by chance any more than an airplane could form simply by sticking all its parts into a huge bag and shaking it for a few millennia. What my friends don't seem to consider is that individual atoms are "compelled" to come together to form intricate compounds as a result of their electronic properties and attraction to form stable chemical bonds. Complex biomolecules do form spontaneously. Over the last forty years, radio astronomers have scanned dust clouds in outer space at infrared and microwave wavelengths, and identified more than sixty organic (carbon-containing) molecules including alcohol, ether [$(CH_3CH_2)_2O$], acetylene (C_2H_2), formaldehyde (H_2CO), and cyanodecapentyne (a 13-atom molecule). The atmosphere of Titan, the largest moon of Saturn, is known to contain at least six hydrocarbons (ethane, propane, acetylene, ethylene, diacetylene, and methylacetylene, i.e., $CH_3C \cdot CH$, where the dot symbol represents a triple bond), three nitrogen compounds (hydrogen cyanide, cyanoacetylene or C_3HN, and cyanogen), and two oxygen compounds (carbon dioxide and carbon monoxide). Titan's atmosphere almost certainly contains other more complex substances.

In comets, researchers have found many possible precursors of life such as methyl cyanide and hydrogen cyanide. In 1986, scientists discovered more than thirty organic (carbon-containing) molecules in Halley's Comet using dust-particle impact spectrometers on space crafts. The organic molecules are relevant to life and include pyrimidines and purines, ringlike molecules that are necessary for the message in our genetic codes.

We also know that sugars and amino acids are easy to create just by shining ultraviolet light in a flask containing a mixture of carbon dioxide, ammonia, and water vapor—a combination that may resemble Earth's primitive atmosphere billions of years ago. Many scientists suggest that our original atmosphere lacked oxygen as evidenced today by certain primitive forms of bacteria that are killed by oxygen, such as those causing gas gangrene or tetanus.

If our flask of gases is also rich in methane and hydrogen, as is the atmosphere of Jupiter and Saturn, and an electrical spark is shot through it for a week, a wonderful array of life chemicals is produced. The water in the flask actually turns deep red and contains: alanine and glycine (amino acids), lactic acid, acetic acid, urea, formic acid, glycolic acid, and more. Other gas experiments have produced components of nucleic acids. Merely shining ultraviolet light onto formaldehyde, a molecule assumed to have been generated in the primordial atmosphere, produces ribose and deoxyribose—sugars of RNA and DNA. Researchers have also easily produced ATP, the energy molecule of all life forms. Although the primitive atmosphere may have had much less hydrogen and more carbon dioxide than in these experiments, and was thus a more oxidizing atmosphere, such oxidizing atmospheres also easily yield the chemicals of life.

I find it quite exciting that out of all combinations of atoms that *might* have been produced by scientists randomly mixing simple molecules like methane, carbon dioxide, ammonia, and water—the ones most *easily* produced are life's building blocks such as amino acids, sugars, fatty acids, purines, and pyrimidines.[6]

What these experiments suggest is that no extraordinary circumstances were needed to catalyze the diversity of biomolecules that eventually led to all the life we see around us. Moreover, if life on Earth could evolve from ordinary processes, then it is likely that humans are not alone in our own solar system. Other life forms, perhaps quite primitive, should exist.

The ease with which the basic building blocks of life are produced in simple experiments once prompted Nobel-Prize-winner Melvin Calvin to write, "We can assert with some degree of scientific confidence that cellular life as we know it on the surface of the earth does exist in some millions of other sites in the universe."[7] Nobel-Prize-winner Christian de Duve wrote, "Life belongs to the very fabric of the universe. Were it not for an obligatory manifestation of the combinatorial properties of matter, it could not possibly have arisen naturally."[8]

Before the early 1800s, many scientists believed that chemicals produced by and within living creatures—organic compounds—were substantially different from chemicals derived from minerals and salts, and that mere humans would never be able to create organic compounds from inorganic ones without the help of living organisms. However, the German chemist Friedrich Wöhler firmly put this religious thinking to rest when he synthesized urea in 1828 from ammonium cyanate, which is generally considered to be inorganic reactant.

Monkeys Typing The Bible

Creationists sometimes argue that the second law of thermodynamics drives nature to disorder and chaos, and, as a result, the intricate and ordered patterns of life could not have arisen spontaneously without an intelligent designer. But nothing could be further from the truth. We don't need to invoke an intelligent designer whenever we see the ordered snowflake. The second law of thermodynamics does not preclude pockets of order from arising naturally, especially when we consider that life-giving energy is continually pumped into our world in the form of sunlight. The sun drives photosynthesis, which has enabled complex life to flourish.

Recall the creationist argument that evolution is as preposterous as the idea of shaking airplane parts in a huge bag and expecting an airplane to emerge. However, this is a poor analogy for evolution that harnesses nonrandom change by preserving useful features and eliminating nonadaptive ones. Natural selection pushes evolution in particular directions that contribute to the survival of a creature's offspring.

As a useful way to see how structure can arise by eliminating nonadaptive features, consider the phrase, "In the beginning, God created the heavens and the earth." How long would it take a monkey to type this phrase, if it typed random letters? How long would it take the same monkey to type this phrase if the monkey generated letters randomly while preserving the position of individual letters that happened to be correctly placed—which, in effect, selects for output that is more like the biblical phrase?

Let's tackle the first question. The phrase contains fifty-six letters (counting spaces and the period at the end). If the probability of hitting the correct key on the typewriter is $1/n$, where n is the number of possible keys, then we have the probability of typing the entire phrase correctly as $1/n^{56}$.

To compute this quantity, let's ignore the fact that the monkey would have to hold the shift key to produce capital letters and just assume that the monkey can hit any of $n = 93$ potential characters that include

abcdefghijklmnopqrstuvwxyz
ABCDEFGHIJKLMNOPQRSTUVWXYZ
1234567890`~!@#$%^&*()_+=/,.[]{}|\';"

Thus, the probability of the monkey correctly typing fifty-six consecutive characters in the target phrase is, on average, $1/93^{56}$, which means that the monkey would have to try more than 10^{100} times, on average, before getting it right! Stated with more mathematical precision, the monkey would have to try 1.19×10^{110} times in order to have a 50-50 chance of typing the phrase correctly. If the monkey pressed one key per second, he'd be typing for well over the current age of the universe. Also, if every particle in the universe were using a virtual typewriter, and could type one letter each microsecond, it would take the age of the universe to get a 50-50 chance of getting the correct sentence![9]

However, in the second case in which we save characters that are typed correctly, the monkey will obviously require many fewer keystrokes. Mathematical analysis reveals that the monkey, after only 407 trials, would have a 50-50 chance that the correct sentence has been typed![10] This crudely illustrates how evolution can produce remarkable results when harnessing nonrandom changes by preserving useful features and eliminating nonadaptive ones.

Robert Ardrey and the Killer Ape

Available evidence suggests that humans evolved from apelike ancestors over the millennia. A rather poetic take on human evolution was provided by Robert Ardrey (1908–1980), an anthropologist interested in human behavior. The chameleonic Ardrey was also a Hollywood screenwriter, having written such notable movies as *Khartoum, The Four Horsemen of Apocalypse, The Wonderful Country, Madame Bovary, The Secret Garden,* and *The Three Musketeers.*

Ardrey returned to anthropology after his movie writing career—and his numerous books often focused on behavioral evolution and on how humans interacted with one another. Ardrey's most famous book is *African Genesis,* which discusses the role that aggression plays in human evolution and society. Ardley speculates on our origins in *African Genesis:*

Man had emerged from the anthropoid background for one reason only: because he was a killer. Long ago, perhaps many millions of years ago, a line of killer apes branched off from the non-aggressive primate background. For reasons of environmental necessity, the line adopted the predatory way. For reasons of predatory necessity, the line advanced. . . . And lacking fighting teeth or claws, we took recourse by necessity to the weapon. . . . But the use of the weapon meant new and multiplying demands on the nervous system for the coordination of muscle and touch and sight. And so at last came the enlarged brain; so at last came man.[11]

These days, Ardrey's killer ape hypothesis is not without controversy and is sometimes viewed as painting human ancestors a little too violently. The protohumans were hunters, but it is not clear that they were any more violent than any other predatory animal and may have murdered their own kinds less than other animals. Chimps do kill each other, and humans competed successfully because their cerebrums allowed them to generate ideas, to plan, to make associations, to make tools, and harness fire. Nonetheless, Ardrey's main point seems valid: The struggle for survival is the contest our ancestors entered, and you are here today, able to read this book, because we had a little luck and also played the contest well enough—better than some of our other ancestors and relatives competing for the same niche.

Perhaps Ardrey's observations of human violence began when he was a small boy in Chicago. While his family attended services at a neighborhood church, he hung out with the younger boys in the church basement to fight. He writes,

A new member or two would be initiated and if injured seriously, helped home to his mother. There would be a short prayer and a shorter benediction. And we would turn out all the lights and in total darkness hit each other with chairs. It was my Sunday-school class in Chicago, I believe, that prepared me for African anthropology.[12]

Ardrey does admit that human survival has been advanced by a violence-sublimating "super-territorial institution" called civilization. But rather than see this ability to live in large cooperative groups as a shining glory of humankind, he likens it to a crumbling house:

> [Civilization] is a jerry-built structure, and a more unattractive edifice could scarcely be imagined. Its grayness is appalling. Its walls are cracked and eggshell thin. Its foundations are shallow, its antiquity slight. No bands boom, no flags fly, no glamorous symbols invoke our nostalgic hearts. Yet however humiliating the path may be, man beset by anarchy, banditry, chaos and extinction must at last resort turn to that chamber of horrors, human enlightenment. For he has nowhere else to turn.[13]

The Religious Implications of Mosquitoes

A few of my creationist friends admit that we see adaptations emerging in many modern species of bacteria, fruit files, fish, and squirrels, but they counter that these impressive adaptations occur within a species—and we have never seen one species transform into another, as would be required by evolution. But why focus so strongly on the notion of species when thinking about genetic and body modification through time? Robert Pennock writes in *Discover* magazine, "There is no essential difference between adaptations 'within' a species and those that create new species, so if evolution has the power to do the one, it requires no extrapolation to get the other."[14] In some ways, it's more impressive to me that we have been able to evolve a Chihuahua from a wolf (a change within a species, that every creationist seems to accept) compared to a tiny evolutionary change that may be required to turn one species of beetle or plant into another—a change that many creationists can't accept but that yields creatures so close in appearance that they cannot be distinguished, even by experts, without intensive molecular examination.

One key component of evolution is a change in the gene pool of a population over time. For example, insect populations develop resistance to pesticides over the period of a few years. Thus, most creationists recognize and accept that evolution of gene pools is a fact. Again, if gene pool changes can happen so quickly and dramatically, perhaps the change in a gene pool that transforms one species into another species, which some creationists see as an impossible boundary to penetrate, is not insurmountable after all.

In any case, we have indeed seen species transform into other species in just the last hundred years. New species have arisen before our very eyes and seem to be in the process of emerging all over the globe. Most of the time, this recent speciation has been confirmed by observing organisms living only in environments that did not exist a few centuries ago. For example, a species of the *Mimulus* flowering plant has recently evolved to live in soils high in copper, and the plant only exists on the edges of a copper mine that did not exist before 1859. The new species has smaller flowers and is more branched than the progenitor forms. Unlike typical *Mimulus*, which insects pollinate, these new plants are self-fertilizing.

Is the new *Mimulus* plant truly a different species? Botanist Mark Macnair, now a professor at the University of Exeter in England, considers the copper-loving *Mimulus* so different from the ordinary ones that they are a separate species, which is now called *Mimulus cupriphilus*, the copper-loving monkeyflower. As a self-pollinator blooming a month or so earlier than the normal *Mimulus guttatus, cupriphilus* is reproductively isolated from *guttatus* in time. Macnair suggests that *cupriphilus* has four genetic systems that are different from *guttatus*, and that the key mutation is the one for earlier flowering.[15]

Many new species of plants have arisen via polyploidy, a condition common in plants that results when the number of chromosomes are doubled. For example, in the 1940s, a new, fertile species of primrose plant (*Primula kewensis*) was produced through chromosome doubling in a hybrid of two primrose species. *Kewensis* cannot breed with either of its parent species, *Primula verticillata* or *Primula florbunda*, but it produces healthy offspring from its own kind. Scientists estimate that over half of the angiosperms (plants with hard seeds) are the product of natural polyploid evolution.

Here's an example of how you and I could foster the emergence of new species in our own basement laboratories. In 1981, Richard Halliburton and G. Gall studied a population of flour beetles collected in Davis, California. In each generation they separated the eight lightest-weight and the eight heaviest pupae of each sex. When beetles emerged from the pupae, they were placed together and allowed to mate for twenty-four hours. Pupae that developed from these eggs were then segregated according to weight, and the process repeated for fifteen generations. As a result, the researchers created a race of heavy and light beetles, and they discovered—not too surprisingly—"positive assortative mating on the basis of size."[16] "Positive assortative" mating refers to the fact that the heavy beetles tended to mate only with heavy beetles, and the light beetles tended to mate with the light beetles. If Halliburton and Gall had the facilities and patience to carry on the experiment for a decade, it's not far-fetched to think that their two races would have developed into different species that could not have mated with one another.

More exciting than the beetles and plants is the discovery of a potentially new species of mosquito, *Culex molestus*. This mosquito lives exclusively in the tunnels of the London subway system, and may have evolved from the *Culex pipiens* mosquito in just a few years. In the late 1990s, the elegant Katherine Byrne, now a geneticist at the Institute of Zoology in London, prowled the subway systems after midnight, gazing at pungent pools of stagnant water that ranged from "pretty nasty" to "hideous," in her quest for hot spots for mosquito-breeding sites. Byrne discovered that although *Culex pipiens* (aboveground mosquitoes) and *Culex molestus* (subterranean mosquitoes) look identical, their behaviors are quite different, and they can't breed to produce offspring. When Byrne crossbred the *pipiens* and *molestus*, none produced viable eggs, which suggests that *molestus* is reproductively isolated from *pipiens*, the traditional signature of a new species.

Byrne and Richard Nichols, of Queen Mary and Westfield College at the University of London, soon confirmed that different colonies of the underground mosquitoes were more genetically similar to one another than to their aboveground counterparts.[17] Fecund enclaves of subterranean creatures were being established by other underground mosquitoes, not by mosquitoes that lived in open air. Also, the behaviors of the two mosquitoes are quite different. *Pipiens* hibernates in winter, and *molestus* cannot survive the cold. *Pipiens* must swarm in the open before mating, whereas *molestus* thrives in confined spaces.

At least three genetically distinct subvarieties of *molestus* have been discovered, each one unique to a different subway line. The researchers believe that their studies are most consistent with *molestus* populations evolving from local populations of *pipiens*, and that the reproductively isolated London Underground mosquitoes arose within just a few hundred generations.

If some religious Christians say evolution is not possible, and we are watching mosquitoes and plants in the process of forming new species, what are the religious implications of such observations? How can people deny evolution, if we can see it happening, or on the threshold of happening, all around us?

The Omphalosian View of Reality

My favorite subject in high school was biology. I was so obsessed with biological knowledge that I also took extra high school classes in invertebrate zoology, vertebrate zoology, and anatomy and physiology—classes that are probably not often taught in high schools today. During high school summers, I went to schools in oceanology and environmental science. I got an 800, a perfect score, on the high school biology achievement test. I had human anatomy posters hanging in my bedroom and wore T-shirts featuring dissected frogs or the human circulatory system. I guess I really was a biology nerd!

As Mr. Galbraith, the biology teacher, explained to our class: "Evolution is a fact. Living things change, and this change occurs over long periods of time." Years later, I still remember his words, because they made a lasting impression, and I don't think most high school teachers could get away with saying something quite so strong today, especially in some of the states in middle America. Nevertheless, the fossil record is clear: species have changed over time. Trilobites existed long before mammals and are now extinct. The exact mechanism of evolution is a theory based on fossil and other data.

Of course, some creationists would not accept my trilobite argument and instead advance the Omphalos argument, first expounded in Philip Henry Gosse's 1857 book titled *Omphalos: An Attempt to Untie the Geological Knot.* The omphalosians today argue that that the universe and the Earth were created in the last few thousand years but given the *appearance* of age. In the early 1960s, John C. Whitcomb and Henry Morris published *The Genesis Flood: The Biblical Record and Its Scientific Implications.* In the book, they argue that Earth's original soils were created with the *illusion* of being much older than they actually are. The name *Omphalos* derives from the Greek word for belly button and the earlier Christian debates as to why Adam had a navel if he was never born with an umbilical cord. For Adam, God also gave the appearance of some past event or history that never existed. Gosse called these events prochronic ("outside time") and reasoned that the history of the Earth and civilizations contained numerous prochronic events.

Arkansas Act 590

Although proponents of "intelligent design" indicate that the cell, brain, or eye is too complex to have emerged through evolution over millions of years, this is not supported by fact. Given time, wolves can turn into poodles, and the lovely complexity and symmetry of a snowflake arises spontaneously from the laws of chemistry. Additionally, just because something has the appearance of design does not mean it is designed. The sun appears to literally rise in the sky in the morning and plummet in the evening.

The beautiful, symmetrical snowflake may appear to be designed, but we need not invoke an intelligent designer for the snowflake. We certainly don't need laws that force public schools to give equal treatment to intelligent design and physics for the creation of snowflakes or Buckyballs—beautiful soccerball-shaped molecules that form naturally in interstellar space. In any case, apparent complexity need not be the primary marker for design. After all, a perfect cubical stone, sitting on a beach of stones with complicated shapes, is more likely to be a product of design.

In the early 1980s, Arkansas Act 590 was enacted into law. It required "balanced treatment of creation science and evolution science in public schools." However, federal judge, William R. Overton of Arkansas, ruled that creation science conveys "an inescapable religiosity" and therefore is unconstitutional. He also said that it was not a science because the "essential characteristics" of a scientific explanation are:

1. It is guided by natural law.
2. It has to be explanatory by reference to natural law.
3. It is testable against the empirical world.
4. Its conclusions are tentative.
5. It is falsifiable.
(McLean v. Arkansas, US District Court, 1982)

Overton concluded that creation science fails to meet these essential characteristics. Regarding point 5, evolution is actually falsifiable. For example, if we frequently found an ape in the same geological layer and with the same age as a trilobite, as determined by scientific dating methods, this would be a fairly good refutation of evolution. On the other hand, how would a creationist falsify the notion of creationism? Additionally, evolution can be used to make predictions in the sense that it foretells what will be found when excavating fossils. It also predicts the kinds of results that happen in fast-evolving species like bacteria. The prediction that penicillin-immune bacteria would develop is a very good example of the predictive power of evolution.

Much of the evidence for evolution is indirect and not obtained by laboratory experiment. However, this should not make evolution any less scientific than other experimental fields of science. For example, evidence in the fields of geology and cosmology is often obtained indirectly and not through laboratory experiments. But this doesn't mean that our ideas about the geological, astrophysical, or cosmological phenomena are mere unsubstantiated guesses.

Who Is the Designer?

Whenever I listen to proponents of intelligent design on TV, they don't seem to speculate much about the nature of the intelligent designer, and I'm not sure why. They also don't theorize about how the intelligent designer accomplished the task. This seems odd, because whenever one proposes a theory, these questions would seem to be foremost in a person's mind, especially for scientists. For example, if I believe the Earth to be warming, I would be curious about knowing how and why. As another example, if fossils suggest that the *Lystrosaurus*, a mammal-like reptile, roamed the countryside of Africa, Madagascar, India, and Antarctica, I would wonder if there was once a physical connection between these regions. Perhaps any scientist would wonder how this happened and imagine steps needed to provide an explanation. The overwhelming structural, paleomagnetic, fossil, and climatic evidence supports the theory that Madagascar was once attached to these lands, before the continents drifted apart. As Madagascar broke away from its connection to its neighbors, the long period of geographic isolation allowed for a unique brand of speciation to occur throughout the island.

When people speak of an intelligent designer, what are they really saying? As we learn more and more about science, the intelligent designer seems to have less to do. Michael Shermer, author of *Science Friction*, notes:

> By a different name for a different time, the Intelligent Designer (God) used to control the weather, but now that we have a science of meteorology, the Intelligent Designer has moved on to more obdurate problems, such as the origins of DNA or the evolution of cellular structures such as the flagellum. Once these problems are mastered, then the Intelligent Designers will presumably find even more intractable conundrums.[18]

Shermer is suggesting a diminishing role for the intelligent designer in the twenty-first century. According to supporters of intelligent design, the designer is no longer needed to make lightning but instead is required to make small objects inside cells. But when scientists learn more about the cells, a time will come when no one suggests that a designer is needed to make the cell component. Just because we don't understand something now does not mean that we won't be able to explain it scientifically in the future. For example, all sorts of current mysteries on earth might require someone to say God or some intelligent designer was responsible. However, do we need to suggest that some advanced force was needed to move hundreds of giant stone statues erected on the coast of Easter Island? Or is it more logical to say that some mysteries exist, but they will cease being mysteries once we learn more? Shermer writes:

> If Intelligent Design theory is really a science [as its proponents claim], then the burden is on them to discover the mechanisms used by the Intelligent Designer. And if those mechanisms turn out to be natural forces, then no supernatural forces are necessary, and proponents of Intelligent Design can simply change their name to scientists.[19]

God's Laryngeal Nerve

If we look closely at biological structures, we find apparent clumsiness and inefficiencies in the so-called intelligent design. As just one example, *New York Times* columnist Jim Holt writes about a long nerve in mammals in "Unintelligent Design":

> The recurrent laryngeal nerve in mammals does not go directly from the cranium to the larynx, the way any competent engineer would have arranged it. Instead, it extends down the neck to the chest, loops around a lung ligament and then runs back up the neck to the larynx. In a giraffe, that means a 20-foot length of nerve where 1 foot would have done. If this is evidence of design, it would seem to be of the unintelligent variety.[20]

In other words, creatures have many inefficient structures in their bodies, which suggests that an advanced force did not directly design these structures. Aside from the "inefficient" laryngeal nerve, we find nonfunctional wings in kiwi birds, useless portions of eyes in some creatures confined to caves, and the infection-prone appendix in humans. Those who believe that God designed humans may be shocked to learn that less than one-third of human conceptions culminate in live births. This means that more than two-thirds end prematurely. So, many embryos die! Holt concludes that, "nature appears to be an avid abortionist, which ought to trouble Christians who believe in both original sin and the doctrine that a human being equipped with a soul comes into existence at conception."[21] If these naturally aborted embryos have the taint of original sin, do they stay in a kind of limbo and not enter heaven? Does more than two-thirds of humanity cry out from the abyss for their omniscient and intelligent creator?

Humans and Aliens Will be Immortal

As I ponder the kinds of evolutionary processes that may take place on Earth and on other planets, I also enjoy thinking about the typical life spans of space-faring aliens, if such aliens exist. American astronomer and astrophysicist Frank Drake believes that any intelligent aliens we may someday encounter will be immortal. In 1976, he wrote, "It has been said that when we first discover other civilizations in space, we will be the dumbest of them all. This is true, but more that than, we will probably be the only mortal civilization."

Space-faring aliens will probably live for centuries because they will have solved the mysteries of aging or can repair any damage that might be caused by aging. Similarly, humans will soon achieve biological immortality for the same reason. Immortality is not such a rare thing—many creatures on Earth are virtually immortal. As just one example, consider desert creosote plants in Southwest California, some estimated to be over eleven thousand years old. Lichens can live just as long. In 1997, scientists in Tasmania discovered one of the world's oldest living plants, a forty-three-thousand-year-old *Lomatia tasmanica*, or King's Holly.

Of course, philosophers may say that even if one's body could survive indefinitely, the person that I am would not. All of us are changed by our experiences—and these changes are usually gradual, which means that I am nearly the same person that I was a year ago. According to my colleague and futurist Chuck Gaydos, if my body were to survive continuously for a million years, gradual mental changes would accumulate, and an entirely different person would eventually inhabit the body. Thoughts, emotions, and outlooks would compete in my skull like creatures competing for dominance in the process of biological evolution. The million-year-old person—this orchidaceous Ozymandias—would be nothing like me. I would no longer exist. There would be no moment of death at which I had ceased to exist, but I would slowly fade away over the millennia, like a sugar cube dissolving in water, like a sand castle being transformed by an ocean of time.

Abortion
In this century, when humans become virtually immortal as a result of biological manipulations, women will be able to ovulate or delay

ovulation at will, and men will have the choice of producing viable or nonviable sperm. According to Chuck Gaydos, accidental pregnancies and abortions will rarely occur in societies that have achieved immorality. If nanotechnology and nanorobots exist that repair damage from aging, this technology might even be used to prevent conception without the appropriate coded signal from a family's, or government's, master computer.

Returning to the current time, all of my creationist friends who reject evolution also believe abortion is equivalent to murder and should be illegal. In *Sex, Drugs, Einstein and Elves*, I give various reasons why it makes no sense to consider a fertilized human egg (i.e., a zygote) as fully human, and no sense to equate killing the zygote with murder. It always amazes me that those women who consider a fertilized egg fully human still find it acceptable to take the standard birth control, which kills millions.

Many women believe abortion to be murder, but they are speechless when I ask how much jail time a woman should serve when using the ordinary birth control pill. A significant number of fertilized eggs are aborted when women take the pill. Scientific papers suggest that escape ovulation occurs in 4 to 15 percent of all cycles for women taking birth control pills. The term "escape ovulation" refers to the release of eggs by a woman's ovaries even while she is taking the pill, and this occurs in millions of women each year. Once an egg has been released via ovulation, a woman can become pregnant. Physicians confirm that the pill, IUD, Depo-Provera, and Norplant can cause abortions of fertilized eggs.[22]

A. C. Grayling, professor of philosophy at Birbeck College, University of London, also wonders why we attribute such value to the fertilized egg. He asks where we should draw the ethical line:

> To draw it at the moment a zygote is formed rather than at the point where a fetus becomes independently viable—from where

something really can "become a baby"—is to ignore the fact that nature itself is profligate with the zygote, the morula, the blasto-cyst, the embryo, the fetus, voiding itself of any it is not satisfied with, in numbers unimaginable to the moral sentimentalist for whom the mere existence of life rather than its value—its quan-tity, not quality—is what matters most.[23]

Recent studies by Luc Bovens of the London School of Economics show that women who practice the rhythm method as a form of birth control condoned by the Catholic Church actually kill more embryos than women who use other contraceptive methods! Even when women attempt to avoid pregnancy by avoiding sex during the woman's fertile period, some conceptions do take place outside the fertile period, and these fertilizations have a high change of leading to failed embryos. Writing in the 2006 *Journal of Medical Ethics* (Volume 32), Bevans notes that "millions of rhythm method cycles per year globally depend for their success on massive embryonic death. . . . Even a policy of prac-ticing condom usage and having an abortion in case of failure would cause less embryonic deaths than the rhythm method."

Hepatoscopy and Beyond

I'd like to end this chapter by returning our attention to the liver—although I'm sure that Sylvia Weinberger, our Queen of Chopped Liver, would have preferred to skip this lurid section if she were still alive. For centuries, mystics, madmen, and the mythopoeic have gazed at livers in attempts to predict the future.

Entrail reading, or divination by studying animal organs, was one of the most common forms of divination in ancient times. By 2000 BC, the art of divination by the study of sheep livers was pop-ular in Mesopotamia. The ancient Etruscans, Greeks, and Romans also had a passion for sheep guts. For example, Etruscans thought they could read messages in the surface of the liver. They also studied liver folds and veins, along with other features of sheep spleens, lungs, and hearts.

Let's consider hepatoscopy or hepatomancy, a practice that focused exclusively on examination of the liver of sacrificed animals. The Babylonians were famous for hepatoscopy and considered the liver the source of the blood and hence the source of life. The *baru* or priest was specially trained to interpret markings on the livers. Many armies actually brought along a baru to recommend actions before battles. Private citizens also employed barus to perform hepatoscopy.

The prophet Ezekiel in the Old Testament refers to hepatoscopy in 21:21: "For the king of Babylon stood at the parting of the way, at the head of the two ways, to use divination: he made his arrows bright, he consulted with images, he looked in the liver." The hepatomancy system of the Sumerians required looking for thousands of possible variations in the size, shape, texture, and health of the liver in order to foretell the future.

The Greeks practiced entrail reading as part of their religion, and the readings played a role in major Greek events—the building of the Parthenon temple, the battles of the Athenian naval strategist Themistocles, or the decision to enter the Lamian war (323–322 BC)—the conflict in which Athenian independence was lost despite efforts to be free of Macedonian domination after the death of Alexander the Great.

The Etruscans of western-central Italy couldn't get enough of entrail reading, and it seems that they wouldn't even make minor decisions without consulting animal guts. The Etruscans had ruled Rome until about 500 BC, and their entrail readers, called haruspices, performed their art for the Romans. The Roman statesman Cicero said that, "the whole Etruscan nation was stark mad on the subject of entrails."[24]

The haruspices (hah-RUS-puh-seez) continued their practice, also called haruspimancy or haruspication, until the fall of the Roman civilization. I have gazed at an Etruscan bronze model of a sheep's liver found in Piacenza, Italy. The shape from the second century BC is covered with mysterious writing, and subdivided into numerous different sections with crisscrossing lines. Each section seems to contain the name of an Etruscan god. Today, veterinarians and scientists feel that wrinkles in a sheep's liver are caused by the pressure of surrounding organs and by various environmental and genetic factors. In humans, physicians mostly ignore these wrinkles because they have no effect on an individual's well-being. The ancients, however, were fascinated by two specific liver wrinkles that they believed indicated the "divine presence" in the sheep.

Other mideastern cultures practiced entrail reading. Babylonians used a grid, somewhat like a checkerboard, to section the liver into fifty-five pieces. Marks in each of the sections gave clues as to the future. Here are some examples to give you a feeling for entrail reading:[25]

- A cross-shaped wrinkle in one particular section indicates that an important person will kill his lord.
- Two wrinkles in another section indicate the traveler will reach his goal.
- If two lines look like fingers on the right of the liver, two important people will rival one another for power.
- If both lungs show redness, there will be a fire.
- If the gall bladder is enclosed in fat, there will be cold weather.
- If the diaphragm clings, there will be divine support.

In ancient Mesopotamia, when the priest examined the entrails, he usually reported on them in a customary order starting with the liver, with the parts taken in a counterclockwise order beginning with the left lobe. Next came the lungs, the breastbone, the stomach, the vertebrae, the spleen, the pancreas, the heart, the kidneys, and the intestines. In many cases there would be a "consistency check" by sacrificing a second animal and making sure the results were compatible with the first reading. The procedure remained essentially the same for a thousand years.[26]

Toward the end of the Old Babylonian period, around 1700 BC, the science of haruspimancy moved out westward to the countries of

Levant (countries along the eastern Mediterranean shores). Models of livers, sometimes with inscriptions, have been found at Ugarit (northern Syria), Megiddo and Hazor (Israel), and Ibla (Sicily). Model lungs have been found at Ugarit and Alalakh (Syria).

The Hittites, ancient peoples of Anatolia (modern-day Turkey), dissected "cave-birds" either to confirm the inspection of sheep entrails or possibly to provide an independent method of divination. (The cave-bird is thought to be a kind of partridge.) Wealthy Hittites often performed multiple divinations. There is one record of multiple inquiries that required thirty-four inspections of sheep entrails and twenty cave-birds. Perhaps such an expensive, extensive set of entrail readings would be used mainly by the royalty for affairs of state and military campaigns. The Hittites also enjoyed watching the victim's movements as it died. I bet you're not going to feel the same the next time you sit down for a dish of fried liver and onions.

The Roman senate often used one haruspex (Latin: *haruga*, a victim; *specere*, to inspect), to monitor another to make certain he was truthfully telling all he read in a liver. After all, Roman leaders wouldn't want to misinterpret or overlook one little line in a liver that could lead to war and the death of thousands. They wouldn't want to trust it all to one conniving, upstart haruspex who despised the empire.

Roman generals had haruspices on their staffs to help make major decisions. If a sacrificial fire was used, the smoke was also studied. In this book's Introduction, I mentioned the butterfly effect in history. Can you imagine how the history of the world would have changed if one tiny line in some randomly chosen sheep liver were placed one inch lower? You might not be here today. A battle strategy and outcome could easily have changed as a result of the entrail examination and therefore altered the entire cascade of history. In a sense, your very existence may have relied on a Roman general, his harsupex, and a piece of meat from a dead sheep. A single line in a liver could change the universe.

Divination was as commonplace in the past as is golfing or going to the mall today. Some of you may recall from your high-school history class, or Shakespeare's *Julius Caesar*, the soothsayer warning the Roman leader, Julius Caesar, to "Beware the Ides of March."[27] The Ides correspond to dates on the Roman calendar, which, in March, fell on the fifteenth of the month.

In 44 BC, on March 15, Caesar was assassinated by a group of nobles in the Senate House. He was stabbed twenty-three times and fell at the foot of Pompey's statue. William Shakespeare's rendition is grounded in fact because there actually was a soothsayer, Vestricius Spurinnia[28], who gave Caesar the warning about the Ides of March after studying patterns in a goat's liver. Spurinnia was an upper-class patrician and high priest, greatly respected in Rome. As a haruspex, he was probably admired during his life.

Spurinnia and other soothsayers of his time not only looked at the liver's shape but also the color: a healthy red shade was good. Each section of the liver, to which they gave such names as the "gate" and the "table," had particular meanings. Some areas were connected with specific regions of the sky, which linked the art of gut reading to astrology.

For centuries, the liver was thought of as the seat of emotion, and perhaps this is why the Romans and others placed such a high premium on liver structure. Other societies believed that the soul resided in the liver, and even in Shakespeare's day there were various allusions to the liver's powers. Consider for example Shakespeare's comedy *Love's Labours Lost*, performed in 1594. In the story, four young men, dedicated to study and the renunciation of women, meet four young women and inevitably abandon their absurd principles. Berwone, a lord attending Ferdinand, the King of Vavarre, believes that the liver is the seat of love and that love can become so intense that it resembles worship of a god:

This is the liver-vein, which makes flesh a deity.
A green goose a goddess: pure, pure idolatry.
God amend us, God amend! We are much out o' the way.[29]

❧ ❧

Moses also took all the fat around the inner parts, the covering of the liver, and both kidneys and their fat, and burned it on the altar.

–Leviticus 8:16

At one of New York's legendary delis, Zabar's, the average salary for a lox slicer with 10 years experience is $60,000 a year.
–"Food Facts and Trivia: Delicatessen"[30]

". . . If you didn't care about your own future, you wouldn't care whether you were alive to see it, and you wouldn't be going through all this nonsense to stay alive."

"I'm a mammal," said Bean. "I try to live forever whether I actually want to or not."

–Orson Scott Card, *Shadow of the Hegemon*

Creationists believe that God designed all life, and that's a somewhat religious idea. But *Intelligent Design theorists* think that at some unspecified times some unnamed superpowerful entity designed life, or maybe just some species, or maybe just some of the stuff in cells. . . . It doesn't get bogged down in the details.

–The Editors of *Scientific American*, April 2005

Experience tells us that the more complex an artifact is, the greater the number of people involved in its design and implementation. The proponents of Intelligent Design need to be aware that . . . they lay the groundwork for multiple designers, in essence *polytheism*.

–Marvin Jacoby, *The New York Times Magazine* (Letters section)[31]

The first eyes appeared about 543 million years ago—the very beginning of the Cambrian period—in a group of trilobites called the Redlichia. Their eyes were compound, similar to those of modern insects, and probably evolved from light-sensitive pits. And their appearance in the fossil record is strikingly sudden—trilobite ancestors from 544 million years ago don't have eyes.

—Graham Lawton, "Life's Greatest Invention," *New Scientist* [32]

If a new fossil discovery neatly bisects a "gap," the creationist will declare that there are now two gaps!

—Richard Dawkins, quoting Michael Shermer, "Creationism: God's Gift to the Ignorant," *The Times* (London)

Do oysters have little bivalve souls? Do they dream briny dreams, scream briny screams?. . . They are alive when shucked right in front of us, their deaths more proximal than those of so many creatures we eat.

—Frank Bruni, "It Died for Us," *The New York Times*

I assert first the paradox that our predatory animal origin represents for mankind its last best hope. . . . We were born of risen apes, not fallen angels, and the apes were armed killers besides. And so what shall we wonder at? Our murders and massacres and missiles and our irreconcilable regiments? For our treaties, whatever they may be worth; our symphonies, however seldom they may be played; our peaceful acres, however frequently they may be converted into battlefields; our dreams, however rarely they may be accomplished. The miracle of man is not how far he has sunk but how magnificently he has risen. We are known among the stars by our poems, not our corpses.

—Robert Ardrey, *African Genesis*

EIGHT
THE WHISPERS OF HISTORY
꙳ 𖤣

In which we encounter mathematical formulas, Srinivasa Ramanujan, Jean-Paul Sartre, large libraries, Malcolm X, dictionaries, wonders of the modern world, Will Durant, the waves of history, knots, Isaac Asimov, James Burke, Konstantin Tsiolkovsky, Aleister Crowley, Jack Parsons, L. Ron Hubbard, Robert Heinlein, Marjorie Cameron, Alvin Toffler, Karl Japsers, the axial age, Marshall McLuhan, Truman Capote, Wilhelm Röntgen, Velcro, Charles Goodyear, Harry Coover's superglue, the mathematics of rapture, the Doomsday Argument, Nick Bostrom, the phalanges of history, and Charles Fourier.

It cannot be that our life is a mere bubble, cast up by eternity to float a moment on its waves and then sink into nothingness. There is a realm where the rainbow never fades, where the stars will be spread out before us like islands that slumber in the ocean, and where the beautiful beings, which now pass before us like shadows, will stay in our presence forever.
—George D. Prentice, *Man's Higher Destiny*, 1860

Wonders of the World
I enjoy Top Ten lists and have been collecting various quirky lists for decades. A few years ago, Nicaragua issued ten postage stamps bearing *Las 10 Formulas Matematicas Que Cambiaron La Faz De La Terra*, that is, "The 10 mathematical formulas that changed the face of the world." Isn't it admirable that a country so respects mathematics that it devotes a postage stamp series to a set of abstract equations?

In addition to scientific or mathematical merit, perhaps the Nicaraguan government considered such practical issues as space limitations so as to avoid long formulas on small stamps. For example, no matter how wonderful we may find the following mathematical relationship, it's unlikely we'll ever see it on a stamp.

$$1 + \frac{1}{1 \cdot 3} + \frac{1}{1 \cdot 3 \cdot 5} + \frac{1}{1 \cdot 3 \cdot 5 \cdot 7} + \ldots + \cfrac{1}{1 + \cfrac{1}{1 + \cfrac{2}{1 + \cfrac{3}{1 + \frac{4}{1 + \ldots}}}}} = \sqrt{\frac{\pi e}{2}}$$

This formula by Indian mathematician Srinivasa Ramanujan (1887–1920) draws a shocking connection between an infinite series (at left) and a continued fraction (middle). Both the series and the continued fraction can be expressed through the famous numerical constants π and e, and the series and the fraction both mysteriously equal $\sqrt{\pi e/2}$.

Here is a list of Nicaragua's postage stamp equations for *Las 10 Formulas Matematicas Que Cambiaron La Faz De La Terra*. Do you recognize several of these formulae?[1]

1. $1 + 1 = 2$

2. $F = Gm_1 m_2 / r^2$

3. $E = mc^2$

4. $e^{\ln N} = N$

5. $a^2 + b^2 = c^2$

6. $S = k \log W$

7. $V = V_c \ln(m_0/m_1)$

8. $\lambda = h/mv$

9. $\nabla^2 E = (Ku/c^2)(\partial^2 E/\partial t^2)$

10. $F_1 x_1 = F_2 x_2$

Many of my previous books are about mathematics, but lately I am trying to expand my repertoire to encompass other areas of knowledge. Every so often, I imagine that I am in the vast library of Jean-Paul Sartre's "Self-Taught Man," a character from his book *Nausea*. The Self-Taught Man loves libraries, and he handles books "like a dog who has found a bone." One day, he decides to explore a particular library by reading every book alphabetically as it is shelved. After seven years, he has reached the Ls:

> Today he has reached "L" . . . He has passed brutally from the study of coleopterae to the quantum theory, from a work on Tamerlaine to a Catholic pamphlet against Darwinism, he had never been disconcerted for an instant. He had read everything; he has stored up in his head most of what anyone knows about parthenogenesis, and half the arguments against vivisection. There is a universe behind and before him. And the day is approaching when closing the last book on the last shelf on the far left: he will say to himself, "Now what?"[2]

In *Nausea*, Sartre is actually making fun of the Self-Taught Man, because the Man is obsessed and probably taking an inefficient route to knowledge and wisdom. However, such epic quests sound exciting to me. Imagine attempting to read the entire *Encyclopaedia Britannica* or the eleven-volume *The Story of Civilization* by Will and Ariel Durant: *Our Oriental Heritage, The Life of Greece, Caesar and Christ, The Age of Faith, The Renaissance, The Reformation, The Age of Reason Begins, The Age of Louis XIV, The Age of Voltaire, Rousseau & Revolution,* and *The Age of Napoleon.*

In Umberto Eco's *The Mysterious Flame of Queen Loana*, the protagonist has lost all memories of his existence, *except* for the books he had read throughout life. In the hope of triggering memories of his actual life, he goes to the country and spends time in his grandfather's attic study. Here, he embarks on a reading spree and focuses on an old illustrated encyclopedia that he finds. He reasons that this would be a useful task if the variety of topics can help dredge up more memories.

Compared to the *Encyclopaedia Britannica* or *The Story of Civilization*, reading an entire unabridged dictionary would seem to be an unproductive and less interesting life task. However, in order to improve his literary skills, Malcolm X decided to focus exclusively on the dictionary while in prison. His epic quest began by his actually copying each page and then memorizing what he had written. For Malcolm X, the dictionary was an abbreviated compendium of human knowledge. He writes: "With every succeeding page, I also learned of people and places and events from history. Actually, the dictionary is like a miniature encyclopedia. . . . I started copying what eventually became the entire dictionary."[3]

As a boy I often visited my father's study to examine his eclectic collection of old books. The book I recall most vividly was called *The Volume Library*, first published in 1911. Its preface stated its goal: "*The Volume Library* is a ready reference for the busy man where he may quickly and easily inform himself about matters of education, history, literature, science, biography, geography, trade, industry, art, and an indefinite number of other equally important subjects. It is as the same time a text where may be studied arithmetic, algebra, geometry, grammar, mythology, hygiene, among a long list of studies. *The Volume Library* is a library in one volume."[4] The edition of the book we had contained all kinds of exotic facts and arcane minutiae. For example, after its list of the "seven wonders of the ancient world" (which are the Pyramids, the Colossus of Rhodes, Diana's Temple at Ephesus, the Lighthouse of Alexandria, the Hanging Gardens at Babylon, the Statue of the Olympian Jove, and the Mausoleum by Artemisia at Halicarnassus), it provided a list of the "seven wonders of the *modern* world":

1. Wireless telegraphy, telephone and radio
2. Automobile
3. Airplane

4. Radium

5. Antitoxin

6. Spectrum analysis

7. X-ray

Which of these would remain on the list today? What would your list contain? Genetic engineering? Computers? Antibiotics? The Internet? Television? According to Paul Niemann, author of the 2004 book *Invention Mysteries*, "the five greatest inventions of all time" are: the harnessing of electricity, the printing press, indoor plumbing, penicillin, and the automobile.

In the sixteenth century, Flemish artist Johannes Stradanus delighted readers with his own personal list of "nine discoveries" that were the most outstanding of his age:[5]

1. The New World

2. The magnetic compass

3. Gunpowder

4. Printing press

5. Mechanical clocks

6. Guaiac wood (erroneously considered as a cure for syphilis)

7. Distillation

8. Silk cultivation

9. Stirrups

My own personal survey of colleagues with respect to the "greatest inventions" yielded these suggestions: wheel, flint axes and knives, agriculture, housing, language, clothes, vaccination, heart-lung machine, anesthesia, antisepsis, saddles, and sailing ships. The scientists surveyed in John Brockman's 2000 book *The Greatest Inventions of the Past 2,000 Years* include the following additional suggestions: computer, television, contraceptive pill, the gun, hay ("Without grass in winter, you could not have horses, and without horses you could not have urban civilization"), chairs, stairs, cities, the human ego, calculus, paper, distributed net-worked intelligence, secularism, the scientific method, the eraser (which lets us "fix" our mistakes), and reading glasses (which have "effectively doubled the working life of anyone who reads or does fine work—and have prevented the world from being ruled by people under forty").

One illustration in the *Volume Library* held a particular fascination for me and kindled an early interest in astronomy, and later science fiction and science in general. The figure shown here clearly illustrated the length of time required to reach the various planets from the Earth if one were to travel in a 100-mile-per-hour airplane.

PLANETS

"Length of Time Required to Reach the Various Planets from Earth via Aeroplane Traveling 100 Miles Per Hour." [From Brubacher, Abram, *The Volume Library* (New York: Educators Association, 1911).]

Many questions came to mind as I studied the figure. Were there undiscovered planets? Did life exist on other worlds? If so, what would their biologies and technologies be like? I was accustomed to science-fiction movies and books, and I imagined all kinds of aliens: heat-resistant spiders that lived in the craters of Mercury, gas-beings flying through the flaming prominences on the Sun, and more conventional life forms residing in Jupiter's water-containing moons. Later, as an adult, I continued to imagine and write about such possibilities. The *Volume Library's* "seven wonders of the ancient world" kick-started my interest in the history of knowledge.

Knowledge and Knots

Many people today have lost interest in the old cultures and ways of life, but we can still enjoy the ancient thinking in the Bible, the *Epic of Gilgamesh, Beowulf,* ancient Greek plays and myths, or even history itself. Historian Will Durant notes,

> Greek civilization is not really dead; only its frame is gone and its habitat has changed and spread; it survives in the memory of the race, and in such abundance that no one life, however full and long, could absorb it all. Homer has more readers now than ever in his own day and land.[6]

The growth of knowledge, ideas, science, and art can be imagined as an ever-expanding whirlpool. From the rim, we look down and see previous knowledge from a new perspective as new theories are formed. Today's conjectures mutate, new theories evolve, and yesterday's impossibilities become part of everyday life. But sometimes I wonder if valuable ideas once existed that we have lost. Durant feels that we are "richer" with ideas and art today than in the past:

> The heritage that we can now more fully transmit is richer than ever before. It is richer than that of Pericles, for it includes all the Greek flowering that followed him; richer than Leonardo's for it

includes him and the Italian Renaissance; richer than Voltaire's, for it embraces all the French Enlightenment. . . . If progress is real. . . it is not because we are born any healthier, better, or wiser than infants were in the past, but because we are born to a richer heritage, born on a higher level of that pedestal which the accumulation of knowledge and art raises as the ground and support of our being.[7]

Each major discovery almost always broadens our horizon of interest, unfolding a new set of questions and possibilities. However, old areas of knowledge often suddenly find startling new applications. For example, ancient math can find undreamed of applications centuries later, and such mathematics has even been used to describe the very fabric of reality. For instance, in 1968, Gabriele Veneziano, a researcher at CERN (a European particle accelerator lab) observed that many properties of the strong nuclear force are perfectly described by the Euler beta-function, an obscure formula devised for purely mathematical reasons two hundred years earlier by Leonhard Euler. In 1970, three physicists, Nambu, Nielsen, and Susskind, published their theory behind the beta-function, which eventually led to modern string theory that posits that all the fundamental particles of the universe consist of tiny strings of energy.

As another example of an old science increasingly finding new applications, consider knots in ropes. As I point out in *The Möbius Strip*, it is not an exaggeration to say that knots have been crucial to the development of civilization, where they have been used to create shelters, to tie clothing, to secure weapons to the body, and to permit the sailing of ships and world exploration. Knot patterns have been engraved on Stone Age burial stones. The Incas used knots as a form of bookkeeping and as "written language" along strings known as *quipu*. Some quipus contain nearly a thousand attached strings, each with clusters of knots to record the flow of goods, labor, and taxes.

The ancient Chinese also used knots for fastening, wrapping, and recording events. A few of today's knots have their genesis in the Middle Ages where they were used with compound pulleys for lifting and pulling loads, which were also usually attached with suitable knots.

Today, knot theory has infiltrated physics, chemistry, biology, engineering, and anatomy, and, in many cases has become so advanced that scientists find it challenging to understand its most profound applications. Modern knot conferences cover an extraordinary array of topics, ranging from conditions that may lead to knotted umbilical cords to the study of loop quantum gravity, a field in which certain space-time braids and knots may create particles like electrons, quarks, and neutrinos. Pick up any modern book on knot theory and you'll deal with a list of impressive sounding phrases like Vassiliev invariants, Conway's skein relations, Knotsevich's theorem, Yang-Baxter quantum equations, Hecke operator algebra, topological quantum field theory (TQFT), Temperley-Lieb algebra, and Artin's relation in braid groups. In a few millennia, humans have transformed knots from artistic engravings on rocks to models of the very fabric of reality.[8]

I think Isaac Asimov had the right idea about the future of knowledge: "I believe that scientific knowledge has fractal properties; that no matter how much we learn, whatever is left, however small it may seem, is just as infinitely complex as the whole was to start with. That, I think, is the secret of the Universe."[9]

Konstantin Tsiolkovsky

In his book *Connections*, James Burke lists eight recent innovations that are the "most influential in structuring our own futures." These are the:

1. Atomic bomb
2. Telephone
3. Computer
4. Production-line system of manufacture
5. Aircraft

6. Plastics

7. Guided rocket

8. Television

Of these eight, the guided rocket is perhaps the most potentially devastating for humanity. The rocket can destroy the Earth or take us to the stars. We've come a long way from the small planes pictured in the *Volume Library*! Rockets were first developed by the Chinese over a thousand years ago, using gunpowder as a propellant and mostly for the purpose of entertainment.

In 1903, high school mathematics teacher Konstantin Tsiolkovsky (1857–1935) published the first technical work on space travel, and the Tsiolkovsky rocket equation, which describes the essence of rocketry in a single simple formula, is named in his honor. The equation clearly shows how a rocket can apply an acceleration to itself by expelling part of its mass with high speed in the opposite direction:

$$\Delta v = v_e \ln \frac{m_0}{m_1}$$

This is equation 7 in the Nicaragua's postage stamp list discussed previously. Here, m_0 is the initial total mass, m_1 the final total mass, and v_e the velocity of the rocket exhaust with respect to the rocket. Δv is the change in the rocket's velocity. Tsiolkovsky's work inspired further research and experimentation in Russia.

Tsiolkovsky had a fascinating life. He was one of seventeen siblings, and, at the age of ten, he lost his hearing as the result of scarlet fever. Once deaf, he couldn't attend school, and he didn't receive any formal education for many years. He taught himself largely from the books in his father's library. During the years 1873 to 1876, he visited Moscow libraries, and was tutored by the eccentric genius Nikolai Fedorovitch Fedorov, a Russian philosopher who was working in a Moscow library at the time.

Jack Parsons: Rocket Man

Today the world remembers Tsiolkovsky as one of the greats in the history of rockets, but the hyperchameleonic man that I associate with rocket science is Jack Parsons. Parsons (1914–1952) was a self-taught rocket fuel expert who helped found the Jet Propulsion Laboratory in Pasadena, the town of his birth. According to counterculture expert and author Richard Metzger,

> When the history of the American space program is finally written, no figure will stand out quite like John Whiteside Parsons. Remarkably handsome, dashing and brilliant, Jack Parsons was one of the founders of experimental rocket research group at the California Institute of Technology, and the group's testing facility would eventually become the Jet Propulsion Laboratory, NASA's rocket design center. Werner von Braun claimed [Parsons] was the true father of the American space program for his contribution to the development of solid rocket fuel.[10]

Rockets and rocket fuel were only two facets of Parsons' many passions. He was deeply into magic, mysticism, and rituals. He was friends with occultist Aleister Crowley and formed a collective that practiced a variety of sexual rituals and animal sacrifice while under the influence of mind-altering drugs. Parsons was not quite as far-out as Crowley, whose wives both went insane, and five mistresses committed suicides. Among Crowley's numerous eccentricities was his habit of defecating on carpets. He claimed his excretions were sacred. He also had his two front teeth filed so that they would be sufficiently sharp to draw blood when he kissed a woman.[11]

Returning our attention to Parsons—he traveled in other impressive circles and was friends with science-fiction writers Robert Heinlein and L. Ron Hubbard, the founder of Scientology. Hubbard actually lived in Parsons's home for a while and "stole" Parsons's girlfriend.

Even as a teenager, Parsons was fascinated by the occult and the concept of Satan. Later in life, he came to believe that magic rituals could tear the very fabric of reality. For example, Parsons and Hubbard performed "The Babalon Working" ceremony, through a ritual practice of masturbation and chanting, in order to remove the boundaries of space and time. Parsons once said of his interest in magic, "It has seemed to me that if I had a genius to found the jet propulsion field in the US, then I should also be able to apply this genius in the magical field."[12]

Although a college dropout, Parsons was eminently scientific and logical—at least early in his life. Later, he believed he had magically conjured his second wife, Candy, through the Babalon Working ritual. Parsons wrote to Crowley on February 23, 1947, "I have my elemental! She turned up one night after the conclusion of the Operation and has been with me since."

Candy (her real name was Marjorie Cameron) was a flamboyant, green-eyed, redheaded twenty-three-year-old who was happy to participate in the Babalon Working to create a moonchild or homunculus to open an interdimensional doorway through which goddess Babalon could enter, and who would speak in the language of Enochian, which we discussed in Chapter 3:

Through the coming of Babalon, Jack thought he would become the antichrist who would bring on the Apocalypse. Richard Metzger writes:

Parsons' Babalon gambit was dazzling to say the least. If the earth must first be covered in evil before the return of the Christ consciousness and the final triumph of good, what better way to hasten the uplifting of humanity than to rip an alchemical hole in the fabric of reality and invite the very spawn of Hell in for a rip-snorting orgy of howling madness.[13]

Marjorie Cameron was one of many unusual people using Parsons's Pasadena home as a meeting place, which had become a virtual boarding house for leading scientists, occultists, science-fiction authors, and cult leaders. Alas, Jack died in 1952 when he dropped a container of mercury fulminate, causing an explosion in his private lab. Controversy has remained over his death, because many wondered how a scientist of his experience could mishandle such a powerful explosive. And why did he have mercury fulminate in his lab at all? Some say that he was working on bizarre experiments, trying to create the homunculus—the tiny artificial man with magic powers of which we previously spoke. Others say he was working on a secret government project. However, a more likely explanation suggests that Parsons was doing movie special effects work at this time, and these chemicals were part of his job. No one knows for sure. What is certain is that the powerful blast blew off his right arm, broke his other arm and both legs, and left a gaping hole in his jaw. He died an hour later at age thirty-seven. When his mother heard the news, she killed herself with an overdose of sleeping pills.

Today, Parsons's memory is honored with a statue at the Jet Propulsion Laboratory. A crater on the dark side of the moon bears his name.

And what ever became of the alluring Marjorie Cameron? Here are some quick highlights of her life.[14] She was born in 1922 in Belle Plain, Iowa. Her grandmother believed Cameron to be a child of the devil because of Cameron's bright red hair. As a teenager, Cameron attempted suicide several times with sleeping pills. In 1943, at age twenty-one, she joined the navy and met Winston Churchill. In 1946, she met Jack Parsons. In 1952, Cameron moved into her friend Renate Druks's Malibu home. In the same year, Druks saw a "bright neon-colored brain with a tail that resembles a spinal column"[15] floating over her bed and was later told that it was Cameron's spirit.

(Note that this image is very close to the image of the floating brain from planet Arous.) In 1953, after Jack's death, Cameron attempted to create a magical child called the "wormwood star" sired by Jack's spirit, and she believed she was pregnant with the magical child.

And you thought *you* had an interesting life.

The Third Wave

In addition to his contributions to the science of rocket fuel, Parsons was also the inventor of JATO (Jet Assisted Take Off) units for aircraft. JATO is a system for ensuring that overloaded planes get into the air by providing additional thrust in the form of small rockets. JATO is one of the technologies that scholars often cite as heralding "The Technological Revolution." Alvin Toffler, author of *The Third Wave*, suggests that while human history is complex, we can observe patterns, or waves, that characterize a phase of civilization:

- *First Wave: The Agriculture Revolution.* (10,000 BC). This wave centers on the rise of crops and farmers. Individual people are "generalists" and perform various different functions in society.
- *Second Wave: The Industrial Revolution* (AD 1750). Tools emerge. Ships, railroads, and automobiles change the developed world. Nuclear families are common. The factory-type education system, the corporation, and mass production emerge.
- *Third Wave: The Technological Revolution* (AD 1950). This wave is dominated by the rise of electronics, information, and computers with accompanying cultural, institutional, and political dislocations. Diverse lifestyles are common. "Adhocracy" communities form on the fly, for example, consider the rapid emergence of a large community working to create the Wikipedia Encyclopedia on the Web.

These three waves still exist and collide in different countries today, especially in impoverished regions of the Earth. Of course, we can debate whether or not these three categorizations of history represent the "best" way to classify the evolution of civilization. For example, my colleague Graham Cleverly points to other waves that we might consider, such as the wave that established major, modern religions or the "navigation wave" that followed the introduction of multimasted square-rigging, which made the great geographical explorations possible and led to European domination of the Americas. Other waves might focus on the Renaissance and Enlightenment, the wave of nation-state building, or Arthur Koestler's "watershed" that divided reliance on religious or other authorities from reliance on observation.

German philosopher Karl Japsers (1883–1969) spoke of the Axial Age when describing the sudden, simultaneous, mysterious appearance of several major world religious and philosophical founders between 800 and 200 BC.[16] The religio-philosophic figures of this tipping point in history include Socrates of Greece, Isaiah of Israel, Zoroaster of Persia, Buddha of India, and Confucius of China. We also have an impressive number of great thinkers in the Axial Age:

- *Ancient Greece*: Homer, Parmenides, Plato, Aristotle, Thucydides, and Archimedes
- *Israel*: Elijah, Isaiah, and Jeremiah (whose messages of ethical monotheism stood out from polytheistic religions of the region and shifted Judaism's course of development)
- *China*: Confucius, Lao Tze, Chuang Tze, and Mo Tze (who taught the concept of universal love in contrast to the Confucian hierarchy of social rites)

Some scholars liken the Axial Age to the axis of a wheel, about which all of human history revolves. The axial age functions as a Viagra pill for ancient religion and philosophy.

Of course, if you look through the list of names associated with the Axial Age, you'll see some gaps. Jesus does not appear in the Axial

THE WHISPERS OF HISTORY

Age, nor does Mohammed. Jaspers places Jesus in the category of "paradigmatic individuals" together with Socrates, Buddha, and Confucius. I do not know if Africa or Native American religions have notable contributors to the Axial Age.

The Axial Age is identified mostly by ideas and philosophies. Marshall McLuhan (1911–1980), Canadian educator, philosopher, and scholar, went beyond philosophies when he gave his list of the "ten most potent extensions of man":[17]

1. Fire
2. Clothing
3. Wheel
4. Lever
5. Phonetic alphabet
6. Sword
7. Print
8. Electric telegraph
9. Electric light
10. Radio/TV ("extensions of the central nervous system" in the same way that the alphabet is an "extension of language")

McLuhan, an English professor at the University of Toronto, was fascinated by the impact of the mass media on society—and was a media darling himself in the '60s and '70s. He enjoyed disarming his critics during debates by saying, "You think my fallacy is all wrong?"

I recently enjoyed watching McLuhan's interview on the *Dick Cavett* TV show, which first aired in December of 1970. During this amazing show, McLuhan meets Truman Capote, and they both ramble on and on about Cavett's trumpet player and how music relates to language. Dick Cavett is nearly speechless as McLuhan and Capote hijack the show with a barrage of one-liners like "Understanding is not a point of view," or "TV drives people inward, but movies drive people outward." I've always enjoyed this McLuhan quotation:

It is not uncommon for people on these trips, especially with new chemical drugs, as opposed to organic ones, to develop the illusion that they are themselves computers. This, of course, is not so much a hallucination as a discovery. The computer is a more sophisticated extension of the human nervous system than ordinary electric relays and circuits.[18]

Serendipity

Some of the greatest discoveries in science have been achieved through serendipity or chance happenings, for example, the invention of Velcro, Teflon, penicillin, nylon, safety glass, sugar substitutes, dynamite, and polyethylene plastics. Even X-rays, so critical to modern medicine and listed earlier in this chapter as one of the "seven wonders of the modern world," came about through chance. In 1895, German physicist Wilhelm Röntgen noticed that a fluorescent screen in his lab glowed when he turned on an electron beam, even when heavy black cardboard should have shielded the screen. He soon determined that invisible rays were passing through the shield and causing the fluorescence. Röntgen, not knowing their nature, called his discovery X-rays.

The phenomena must have been truly amazing. He placed various objects between the tube and the screen, and the screen still glowed. Finally, he put his hand in front of the tube, and saw the silhouette of his bones projected onto the screen. Thus, not only was his discovery of X-rays serendipitous, but immediately after discovering X-rays, he had discovered their most beneficial application to humanity.

Charles Goodyear discovered the process known as vulcanization when he accidentally dropped a piece of rubber on a stove. Similarly, the artificial rubber neoprene was discovered when a lab assistant left a chemical impurity in a test tube. Just after World War II, chemist Harry Coover discovered superglue when a polymer got stuck in his lab instruments. The glue came onto the market in 1958 as "Eastman #910," named after the company for which Coover worked. Use of the superglue during the Vietnam War facilitated the quick closure of wounds.

Superglue (cyanoacrylate) led to another serendipitous find in the late '70s, when Northampton policeman Laurie Wood was mending a photographic tank and suddenly realized that fumes from the glue condensed around his fingerprints. He had accidentally discovered a forensic technique that is now used worldwide to reveal fingerprints.

In 2000, three scientists won the Nobel Prize in Chemistry for their research on how to make plastics conduct electricity. Their work began when a researcher mistranslated instructions and added a chemical a thousand times more concentrated than usual to an experiment.

Sloppy lab work has had profound effects on scientific advancement. For example, Swiss chemist Albert Hoffman got high in 1943 when he touched a speck of LSD, a chemical he had researched for inducing childbirth. Scottish scientist Alexander Flemming was researching the flu in 1928 when he noticed that a mold had contaminated his petri dish and killed the bacteria in it. Legend has it that three leading taste sweeteners all reached human lips by accident when scientists forgot to wash their hands—saccharin (1879), cyclamate (1937), and aspartame (1965). In 1938, chemist Roy Plunkett was searching for a new kind of refrigerant and filled a tank with a gas related to freon. When he opened the tank later, Plunkett discovered a slippery white powder, Teflon, which was subsequently used in the Manhattan Project, for coating cooking pans, and as a component in body parts, because Teflon is one of the few materials that the body doesn't reject.

Viagra began as a potential treatment for angina, and was being tested in clinical trials. It wasn't very effective as an angina treatment, but the researchers soon began to get reports of some unexpected side effects among male users.

By 2020, a significant number of new inventions will be created using "invention machines" that breed solutions to problems by using a computer technique called genetic programming. Already, the creations of such machines are earning patents and outperforming humans in terms of creative inventions. Stanford University professor John Koza, for example, designed an invention machine that consisted of one thousand networked personal computers. By running for several days, the machine generates inventions using algorithms inspired by Darwinian natural selection. In 2005, the machine earned a patent for making factories more efficient, and the patent examiner did not know that the solution was the work of a computer.

Scientists at NASA Ames Research Center have used a similar approach to create a new antenna that initially looked like an ugly mistake, but has since turned out to be far more efficient than a human could devise. The NASA invention machine started with a basic antenna and subsequently made thousands of small, random mutations to its structure. The antenna offspring were tested, and those mutations that produced better antennas were used to create new offspring. After hundreds of generations, the invention machine suggested an antenna that looked like an ugly bent paper clip—yet the serendipitous recombinations performed by the machine led to a creative solution beyond the realm of human innovation.

The Mathematics of the Rapture

As I reflect on the magnificent technological progress humanity has made, I am thankful that I live in an age with antibiotics, high-tech surgery, and even "little" things like air conditioning, Google, and Amazon.com. It must have been dreadful living in past ages without anesthetics or human rights—or when people practiced trepanation (drilling holes in the skull for supposed pain relief). Trepanation was performed worldwide from ancient times until as recently as the American Colonial era.

Have you ever wondered what it might be like to live in the times of Attila the Hun, Genghis Khan, or even George Washington? Given my interest in mathematics, I sometimes imagine history to be a game of sorts, and I contemplate the odds of my being been born in the past versus a million years from now. Will people on earth die out soon, as suggested by many people who believe in the Christian Rapture?

Zany questions like this would seem to be beyond the power of mathematical theory, but these kinds of musings have led some researchers to apply mathematical probabilities to show the human race is doomed to die out very quickly—if you believe the math. Their argument, known as the Doomsday Argument, starts by assuming the opposite, that humanity will survive thousands of years into the future, perhaps spreading through the solar system, perhaps colonizing Mars and the moons of Jupiter and creating vast floating space stations that slowly glide to other stars.

However, if we accept this accelerating growth, then this means that virtually anyone who will ever exist will live in the far future. If this is true, you and I are quite unusual in terms of our existence at this point in time. Journalist Jim Holt, writing in *Lingua Franca*, does the calculations:

> Assume, quite conservatively, that a billion new people will be born every decade until the sun burns out [in five billion years]. That makes a total of 500 quadrillion people. At most, 40 billion people have either lived in the past or are living now. Thus, we would be among the first 0.00001 percent of all members of the human species to exist. Are we really so special?[19]

According to Holt, it is unlikely that we are among these earliest of human beings—if we believe that the human population continues with accelerating growth. On the other hand, it's possible that such exponential growth can't possibly continue, even with our technological progress. Perhaps it is likely that some virus or nuclear catastrophe will wipe us out. Perhaps God will suddenly lift the righteous to heaven and consign the remainder to hell. Perhaps a comet will strike the Earth, or some human-made nanomachine will accidentally escape the lab, multiply, and envelope us in a gray goo. Given that it is very "unlikely" that we are living now in a scenario in which the species continues to exhibit accelerating population growth, this would mean that it is quite likely that you and I are living in the End of Days.

Of course, you may be able to find flaws in this argument, but numerous versions of the Doomsday Argument have also been proposed by different authors, and many papers have been published in support of and in refutation of the argument.

Astrophysicist Brandon Carter introduced scientists to the Doomsday Argument in the 1980s, and it was later promoted by the philosopher John Leslie of the University of Guelph in Ontario. To reiterate, the argument assumes that we shouldn't think we are unusual. The argument also assumes that, like most things in nature, we are in an approximately exponentially growing population that will suddenly end someday—just like the exponential growth of bacteria in a petri dish with limited resources. Because most people will appear just before the end, the end probably isn't more than a few doubling times away from now.

If this still sounds confusing or far-fetched, I asked one of my colleagues to help demonstrate the Doomsday Argument using a computer program. We based our work on some of the arguments of philosopher Nick Bostrom who writes, "Nearly everybody's first reaction is that there must be something wrong with such an argument. Yet despite being subjected to intense scrutiny by a growing number of philosophers, no simple flaw in the argument has been identified."[20]

Let us assume that a train exists on an iron railroad track floating in space. There are one hundred train cars. Each car has a number from 1 to 100 that is scratched onto its exterior.

God says, "I am going to toss a fair coin. If the coin falls heads, I will put you in one of the one hundred cars and create one person in each other car. This means there will be one hundred people. If it falls tails, I will put you in one of the first ten cars, and create one person in each of the other first ten. This is a small population of only ten people."[21]

Now suppose that you are in a plush car with velvet seats and you can poke your head out the window for a second to see the number etched into the car. You discover that you are in car 7. Armed with this extra information, there is a 91 percent probability there are *only* nine other people created!

My colleague Mark Ganson actually performed a computer simulation to verify this. In particular, his simulator gives a probability of 0.909179 after fifteen million trials, which produced 845,000 7s. This also means that there is a 90.9179 percent probability that God's coin came up tails. For those of you who program computers, a snippet of computer code from Ganson's Doomsday simulator is provided in the Notes.[22] Thus, it is less probable for us to be among the first 1 percent of humanity born, than it is for us to be among the first 10 percent. And it is less probable for us to be among the first 10 percent than it is for us to be among the first 20 percent and so on.

We don't even have to run a computer simulation to find an answer. Returning to the cosmic train scenario, the probability that I am in room 7 is $(\frac{1}{2} \times \frac{1}{100}) + (\frac{1}{2} \times \frac{1}{10}) = \frac{11}{200}$. To understand this equation, consider that two scenarios exist in which I find myself in room 7—either I'm one person out of a 100 or I'm one out of 10, both with $\frac{1}{2}$ probability. This yields $\frac{11}{200}$. Next, let's consider what the probability is that I am in room 7 (1 out of 10) *and* we are in the scenario where only 10 people exist (1 out of 2)—yielding $\frac{1}{10} \times \frac{1}{2}$ or $\frac{1}{20}$. Finally, the probability that *if* I am in room 7 *then* there are only 10 people is $(\frac{1}{20})$ divided by $(\frac{11}{200})$ or $\frac{10}{11}$, the same answer we obtained in our computer simulation. In the language of conditional probability, P (10 people given room 7) = P (10 people *and* room 7) / P (room 7).

As Nick Bostrom points out, our finding ourselves in the first ten cars (and that there are only ten people in existence) corresponds to a short lifetime of our species and Doomsday around the corner. Doomsday being far away from now corresponds to one hundred people in the cars. Just as discovering yourself in train car 7 increased the probability of the coin having fallen tails, so finding you are human number 40 billion makes it likely that the humans will face their doom soon rather than later. Exactly how much more likely will depend on the precise numbers we use in our argument. Nick Bostrom says, "After hearing about the Doomsday Argument, many people think they know what is wrong with it. But these objections tend to be mutually incompatible, and often they hinge on some simple misunderstanding. Be sure to read the literature before feeling too confident that you have a refutation."[23]

The Phalanges of History

We've focused mostly on lists of technical innovations in this chapter. However, history is also replete with social innovations that are coupled to technology experiments. My favorite oddball engineer of humanity is French social theorist Charles Fourier (1772–1837) who promoted a utopian vision in which society is restructured, and people live in cooperative groups called "phalanges," each numbering 1620 persons and occupying a common building. Each of Fourier's phalanges was to have a vegetable garden.

Fourier emerged out of the nexus of the Romantic Movement, the French Revolution, and the Industrial Revolution, and he was one of several Utopian Socialists who fervently believed that ideal human societies could be established during his lifetime. He disliked the Industrial Revolution, with its factory system, and considered it a short-lived wave to be surpassed by something grander that would contribute to humanity's sense of bliss and contentment.

Although Fourier was born into an influential family of cloth merchants and spent most of his life engaged in commerce, he never liked his work. If we ignore some of his weird scientific theories, which I'll describe in a moment, he is considered to have been an important person in the history of social theory:

> Fourier's vision, together with his criticism of the existing system, places him as one of the most inspired prophets of 19th century socialism. His remarkable psychological insights, such as his championing of brief spells and variety in work, his quickness to see oppression no matter how veiled, and his penetrating concern

with character formations and problems, links him to modern educational theory, the emancipation of women and even personnel management.[24]

Fourier believed that "unrestrained indulgence of human passion" is the only possible way to happiness. According to the 1911 *Encyclopaedia Britannica*, Fourier suggested that "the institution of marriage, which imposes unnatural bonds on human passion, is of necessity abolished; a new and ingeniously constructed system of license is substituted for it."

According to Fourier, a mysterious transformation would follow the worldwide establishment of his new phalanges society. He expected reality and social structures to morph in incredible ways so that:

- lions would turn into peaceful "anti-lions," several times larger than normal lions, which serve humankind and carry people as passengers
- "anti-sharks" would catch fish for people
- "anti-hippopotami" would pull boats
- the light from the Aurora Borealis would be transformed into dew that makes the North Pole have the climate of sunny Italy
- the oceans of Earth would turn into lemonade

Fourier was mildly obsessed with numbers, which led him to predict that his utopia would last eighty thousand years, eight thousand of them in a particularly harmonious period of time in which:[23]

- every woman would have four lovers or husbands simultaneously
- six moons would orbit the Earth
- the world would contain thirty-seven million poets equal to Homer, thirty-seven million mathematicians equal to Newton, and thirty-seven million dramatists equal to Molière

All sexual activities would be allowed in his phalanges, except those involving force, and his list of permitted activities included: lesbianism, homosexuality, sadism and masochism among consenting partners, sodomy, pederasty, bestiality, and sex between close relatives.

Fourier was also a strong feminist, considering the current role of women in his society as a form of slavery. The existing family structure was partly responsible for the subjugation of women, and the phalanges structure would elevate the status of women.

Fourier's ideas about how quickly people can become alienated from society—and how social interaction and happy workers are essential for a well-functioning and satisfied population—were developed later by economists and philosophers, and had a lasting influence on society. Several phalanges (also called phalanxes) were established in the United States, although none succeeded for long. However, his concepts did encourage the institution of the kibbutz among Zionist settlers in Palestine/Israel.

Let us conclude this chapter with Fourier's modest and shy words from the Introduction to his 1808 book *The Four Movements*, which describes his ideas on scientific and social laws:

> In the course of reading, it should be borne in mind that because the discovery it announces is more important on its own than all the scientific work done since the human race began, civilized people should henceforth concern themselves with one debate only: whether or not [my work is original], for if the answer is affirmative, all economic, moral, and political theories will need to be thrown away, and preparations made for the most astounding, and happiest, event possible on this or any other globe, the transition from social chaos to universal harmony.[25]

A rock pile ceases to be a rock pile the moment a single man con-
templates it, bearing within him the image of a cathedral.
                    —Antoine de Saint-Exupery, *Flight to Arras*

For the salvation of his soul, the Muslim digs a well. It would be
a fine thing if each of us were to leave behind a school, or a well,
or something of the sort, so that life would not pass by and
retreat into eternity without a trace.
—Anton Pavlovich Chekhov (1860–1904), "Nauka," in *Complete
                    Works and Letters in Thirty Volumes*

History is subject to geology. Every day the sea encroaches somewhere upon the land, or the land upon the sea; cities disappear
under the water, and sunken cathedrals ring their melancholy bells.
Mountains rise and fall in the rhythm of emergence and erosion;
rivers swell and flood, or dry up, or change their course; valleys
become deserts, and isthmuses become straits. To the geologic eye
all the surface of the earth is a fluid form, and man moves upon it as
insecurely as Peter walking on the waves to Christ.
                    —Will Durant, *The Lessons of History*

With the publication of his dictionary, Samuel Johnson returned
from his researches into the English language the way an
explorer returns from the North Pole, with a sense of having
seen a terrain that others can see only through his account of
what he found there. . . . It's tempting to think of a lexicographer
in terms of the dictionary he produces. But it's just as interesting
to think of what the dictionary does to the man.
                    —Veryln Klinkenborg, "Johnson's Dictionary,"
                    The New York Times[26]

Turn on, tune in, drop out.

   —Timothy Leary's most famous phrase, created after Marshall McLuhan advised him to come up with "something snappy" to advertise the benefits of LSD

Why not? Why not? Why not?

       —Timothy Leary's dying words

We are amazingly privileged to be born at all and to be granted a few decades—before we die forever—in which we can understand, appreciate and enjoy the universe. And those of us fortunate enough to be living today are even more privileged than those of earlier times. We have the benefit of those earlier centuries of scientific exploration. Through no talent of our own, we have the privilege of knowing far more than past centuries. Aristotle would be blown away by what any schoolchild could tell him today.

       —Richard Dawkins, "The Atheist," Salon.Com[27]

CONCLUSION
❧ ❧

In which we encounter Italo Calvino, *Invisible Cities*, Gilgamesh gardens, New Jerusalem, Calvino strings, and spacetime tangles.

Blessed are the chameleons
For they will let in the light.

In his book *Invisible Cities*, Italian writer Italo Calvino (1932–1985) discusses the inhabitants of a city who decide to connect their homes and apartments with various strings. Calvino writes, "In Ersilia, to establish the relationships that sustain the city's life, the inhabitants stretch strings from the corners of the houses, white or black or gray or black-and-white according to whether they mark a relationship of blood, of trade, authority, [or] agency."[1]

As the days passed, the strings grew so thick, interwoven, and complexly textured that the people could no longer walk through the city nor distinguish all the intricate relationships. "When the strings become so numerous that you can no longer pass among them, the inhabitants leave; the houses are dismantled; only the strings and their supports remain."[2]

Pop culture and today's Internet function a little like the Calvino strings. Innumerable "threads" connect movie stars, scientists, priests, inventors, and composers. Sometimes I imagine drawing strings among all the inhabitants of the planet. The tangle of strings would offer us a glimpse of the invisible connections, the network of relationships that envelope the world like an infinitely complex spider web.

In Calvino's tale of the strings, the inhabitants eventually abandon their town and study it from a distance: "From a mountainside, camping with their household goods, Ersilia's refugees look at the labyrinth of taut strings and poles that rise in the plain. That is the city of Ersilia still, and they are nothing."[3]

The string weavers rebuild Ersilia elsewhere, and they weave a similar pattern of strings that they hope will form a different fabric than had emerged in the first Ersilia. Then they abandon it and take themselves and their houses still farther away. Whenever anyone travels to the territory of Ersilia, they encounter "the ruins of the abandoned cities, without the walls which do not last, without the bones of the dead which the wind rolls away: spider webs of intricate relationships seeking a form."

What if we could somehow stand outside of space and time and construct Calvino strings between people through all time? What if we also connect the strings to mythical characters or characters in novels, or to the extraordinary chameleonic people in this book—to better understand how people and myths affect one another and how they shape the past and present. I would enjoy stretching a string between me and Gilgamesh, colored red because the string connects such distant regions of time and space.

Let us imagine traveling with Gilgamesh through his long dark passage from this world to the world beyond. We finally emerge into the jeweled garden, glistening with bushes and trees made from lapis lazuli, rubies, hematite, and emeralds. Amazingly, this vision recurs throughout literature and civilizations, and represents our need for transcendence. Consider that the Gilgamesh garden is something straight out of *The Wizard of Oz*. It's a vision described by users of the drug DMT. It's even the description of the city of New Jerusalem in the Bible's Revelation 21:

> The wall was made of jasper, and the city of pure gold, as pure as glass. The foundations of the city walls were decorated with every kind of precious stone. The first foundation was jasper, the second sapphire, the third chalcedony, the fourth emerald,

the fifth sardonyx, the sixth carnelian, the seventh chrysolite, the eighth beryl, the ninth topaz, the tenth chrysoprase, the eleventh jacinth, and the twelfth amethyst. The twelve gates were twelve pearls, each gate made of a single pearl. The great street of the city was of pure gold, like transparent glass.

I love this imagery of the afterlife or perhaps of a posthuman world. It uplifts me. When I read my *The DK Illustrated Family Bible*, I feel a sense of mystic transport just looking at the illustrations!

On the other hand, only a few moments ago, I read Greta Christina's essay "Comforting Thoughts about Death that have Nothing to Do with God," and felt depressed. After all, I'm a skeptic and unsure that an afterlife exists. Greta, a freelance writer, notes

> The fact that your life span is an infinitesimally tiny fragment in the life of the universe, that there is, at the very least, a strong possibility that when you die, you disappear completely and for-ever, and that in five hundred years nobody will remember you . . . [this] can make you feel erased, wipe out joy, make your life seem like ashes in your hands.[4]

And then I sigh. It makes me sad to look at my hands, eyes, and the eyes of my family members, and to understand that this will all be dust and ashes. Greta admits that she doesn't know what happens when we die, but she doesn't think this essential mystery really mat-ters. She wants her essay to be upbeat as she reminds us that we should be happy because it is amazing that we even get a *chance* to be alive. We get to be *conscious*. "We get to be connected with each other and with the world, and we get to be aware of that connection and to spend a few years mucking about its possibilities."

I suppose her essay does end on a bright note as she enumerates items that contribute to her happiness, like Shakespeare, sex, five-spice chicken, Thai restaurants, Louis Armstrong, and drifting patterns in the clouds.

In some sense, even the Calvino strings give me a great sense of pleasure. Imagine if we had Calvino strings following us wherever we go, from the moment of birth to the moment of death. Imagine that every molecule in our bodies had its own string. When you think about this more deeply, our Calvino strings never really begin or end. When we die, the Calvino strings of the molecules in our body keep going. When we are born, the Calvino strings of molecules from our mother coalesce into our embryonic form. At no point do Calvino strings break off or appear from nothing.

As we age, the molecules in our bodies are constantly being exchanged with our environment. With every breath, we inhale the Calvino strings of hundreds of millions of atoms of air exhaled weeks ago by someone on the other side of the planet. Thinking at a higher level, our brains and organs are vanishing into thin air, the cells being replaced as quickly as they are destroyed. The entire skin replaces itself every month. Our stomach linings replace themselves every five days. We are always in flux. A year or two from now, a majority of the atoms in our bodies will have been replaced with new ones. We are nothing more than a seething mass of eternal Calvino strings, continuous threads in the fabric of spacetime.

What does it mean that your brain has nothing in common with the brain you had a few years ago? If you are something other than the collection of atoms making up your body, what are you? You are not so much your atoms as you are the pattern in which your atoms are arranged. As we have discussed, some of the atomic patterns in your brain code memories. People are persistent spacetime tangles. It's quite possible that you have an atom of Jesus of Nazareth coursing through your body. Gilgamesh, the historical king who ruled the city of Uruk, is part of your brain or tendons or heart. An atom in your retina may one day be in the tears of a happy lunar princess a hundred years from now. On this subject, English poet John Donne (1572–1631) wrote,

No man is an island, entire of itself. Every man is a piece of the continent, a part of the main. If a clod be washed away by the sea, Europe is the less, as well as if a promontory were, as well as if a manor of thy friends or of thine own were. Any man's death diminishes me, because I am involved in mankind and therefore never send to know for whom the bell tolls. It tolls for thee.[5]

If you were to try to draw a boundary around yourself when viewed as a seething nexus of Calvino strings, you would find the boundary to be completely imaginary. As mathematician Rudy Rucker has noted, "The simple processes of eating and breathing weave all of us together into a vast four-dimensional array. No matter how isolated you may sometimes feel, no matter how lonely, you are never really cut off from the whole."[6] Deepak Chopra, MD, goes further and says, "There is no such thing as a person. A person is the interwoveness of interbeingness and does not have a separate identity."[7] Yet, he also believes that nothing happens to consciousness after we die and likens our connection to this world to talking on the phone. If the phone line is cut, nothing happens to the being who was speaking. Others have likened death to taking off a tight shoe.[8]

I get pleasure by watching chameleons, such as the individuals we discussed in this book. Through their extraordinary stories, we get to share a little of spacetime with people like Capote, Cage, Corman, and the chopped-liver queen. These kinds of people make me smile and help me feel connected to other minds.

Harvard psychologist Stephen Kosslyn has made the remarkable suggestion that "your mind may arise not simply from your own brain but from the brains of other people." He notes that all of us set up "social prosthetic systems," or SPSs, in which we rely on others to "extend our reasoning abilities and help us regulate and constructively employ our emotions." A good marriage often occurs when two people can serve as effective SPSs for each other. In some sense, we "lend" parts of our brains to one another. Kosslyn concludes that your mind arises from the combined activity of your own brain and those of your SPSs. Using this line of reasoning, "one might argue that when your body dies, part of your mind may survive."

Seth Lloyd, in his book *Programming the Universe*, conveys his sadness with respect to physicist Heinz Pagels, who died in an accident while he and Lloyd were mountain climbing. Lloyd seeks solace not in God but in information theory. "We have not entirely lost him," Seth writes. "While he lived, Heinz programmed his own piece of the universe. The resulting computation unfolds in us and around us." Seth likens the universe to a giant computer that feeds on information and generates reality. Our departed loved ones are not gone. Their essence and information is still with us because the cosmic computer has used their bits to define the world we encounter.

When we consider my fascination with the *Epic of Gilgamesh*, it is both amusing and profound that I read some of the text on a computer screen in the twenty-first century. How will people in the next century assimilate *Gilgamesh*? Perhaps *Gilgamesh* will be downloaded directly to your brain's fissure of Rolando, and in a few seconds you would gasp and be just a tad wiser. To make *Gilgamesh* come alive, a computer will tickle your superior temporal gyrus. You'll hear the scorpion beings. You'll see new colors.

Yes, we can be happy that we have the chance to watch the chameleons—and perhaps become chameleons ourselves.

• • •

The TV is on in my living room. I stare into the dangling eyes of the black-and-white brain from planet Arous and open the nearby window on a warm spring day. Outside, bright blue butterflies are flapping their wings, and I hear a whisper of laughter from the wisps and eddies of wind. A bee buzzes on the windowsill; birds chirp; a child giggles; and in the distance, the tintinnabulation of bells on an ice cream truck. The sounds ebb and flow, composing a symphony for chameleons.

☙ ❧

Finishing a book is just like you took a child out in the back yard and shot it.

—Truman Capote

Mr. Asimov once told an interviewer about sadly contemplating death and the end of conscious thought. But, he said, he cheered himself with the thought that "I don't have to worry about that, because there isn't an idea I've ever had that I haven't put down on paper."

—Mervyn Rothstein, "Isaac Asimov, Whose Thoughts and Books Traveled the Universe, Is Dead at 72"[9]

What are superstrings made of? As far as anyone knows they are not made of anything. They are pure mathematical constructs. If superstrings are the end of the line, then everything that exists in our universe, including you and me, is a mathematical construction. As a friend once said, the universe seems to be made of nothing, yet somehow it manages to exist.

—Martin Gardner, "Science and the Unknowable"[10]

[I saw] the blazing eyes of mouse lemurs and lepilemurs among the nighttime trees, and a few chameleons aiming their eyes in opposite directions—as the proverb says, keeping one eye on the future and one on the past.

—Alison Jolly, *Lords & Lemurs*

A human being is a part of the whole, called by us "Universe," a part limited in time and space. He experiences himself, his thoughts and feelings as something separated from the rest, a kind of optical delusion of his consciousness. This delusion is a kind of prison for us, restricting us to our personal desires and to affection for a few persons nearest to us. Our task must be to free ourselves from this prison by widening our circle of compassion to embrace all living creatures and the whole of nature in its beauty. Nobody is able to achieve this completely, but the striving for such achievement is in itself a part of the liberation and a foundation for inner security.

—Albert Einstein[11]

Over the past ten thousand years. . . , we've redeemed ourselves with our accomplishments. We're the ones who came up with the Trevi fountain and Scrabble in Braille and Dr. DeBakey's artificial heart and the touch-tone phone. We [and the chameleons] have made our lives better. *A thousand times better.*

—A. J. Jacobs, *The Know-It-All*

Appendix: Cathedrals of the Mind

This section lists additional quotations that relate to the topics discussed in this book. I think of these quotations as extensions of the thoughts contained in each chapter. They serve as conversation starters and stimulate lateral thinking. Readers may enjoy pondering these quotations as they complete each chapter.

Introduction

> Popular culture is constantly changing. . . . It forms currents and eddies, in the sense that a small group of people will have a strong interest in an area of which the mainstream popular culture is only partially aware.
>
> —"Pop Culture," *Wikipedia, the Free Encyclopedia*

> I don't like this notion of a "great achievement." I know too much history to have delusions as to how long these things last. I once defined literary immortality as "a moment in geological time"—and that's the way it is with books. If you write works of great poetry, they can last for hundreds of years . . . but when you write history, you can be ruined in a few years by some discovery. . . . So we have no notion that we're immortal by any means. We'd be very happy if people still know what our names are when we die.
>
> —Will Durant, *Visit with Will and Ariel Durant*

> Of 150,000 manuscripts submitted to publishers each year, only one in three hundred are sold.
>
> —Catherine Wald, *The Resilient Writer*

Chapter 1: Truman Capote and *The Brain from Planet Arous*

Artist, you are a magician. Art is the great miracle and offers proof of immortality. Who still doubts? Giotto touched the wounds of St. Francis, the Virgin appeared to Fra Angelico, and Rembrandt proved the raising of Lazarus. Absolute refutation of all pedantic sophistries: Moses is doubted, but then Michelangelo comes; Jesus is not recognized, but then Leonardo comes. Everything is profaned, but immutable holy art continues to pray. . . . Relic, invincible banner, almighty art, art-god, I revere you on my knees, you last ray from above shining down on our decay. . . .

—Joséphin Péladan, "Manifesto of the Rosicrucians," 1892

Bobby Fischer is the poster boy for the mad chess genius, a species with a pedigree going back at least to Paul Morphy, who after his triumphal 1858–59 tour of Europe . . . wandered the streets of New Orleans talking to himself. . . . The great Wilhelm Steinitz claimed to have played against God, given him an extra pawn and won.

—Charles Krauthammer, "Did Chess Make Him Crazy?"
Time, 2005

In our culture, we tend to move into cities that push nature away from us. In our mental environment, we do the same thing. Most people live within a very conventionalized set of notions that are deeply imbedded in a larger set of notions. When we go to the physical edges, such as the desert, jungle, and remote and wild nature, and when we go to the mental edges with meditation, dreams, and psychedelics, we discover an extremely rich flora and fauna in the imagination. This realm is ignored because of our tendency to see in words, to build in words, and to turn our backs on the raging ocean of phenomena that would otherwise entirely overwhelm our metaphors.

—Terence McKenna in *Trialogues at the Edge of the West: Chaos, Creativity, and the Resacralization of the World* by Ralph Abraham, Terence McKenna, Rupert Sheldrake

Chapter 2: John Cage and the Zen of Music

In the past, you had to memorize or retain knowledge because there was a cost to finding it. . . . Could you spend time going to the library? . . . Now, what can't you find in 30 seconds or less? We live an open-book-test life that requires a completely different skill set.

—Mark Cuban, "Around the Corner," Interview in *Time*, March 20, 2006

I tried to imagine what it was like for the family of Arnold Schoenberg to appreciate his atonal music, which, to many at the time, sounded like air horns and geese honking, and drove some hearers to riot. And who in Jackson Pollock's family could have foreseen that when he dripped his paints across a canvas on the floor he would become a famous abstract expressionist.

—Cathy Lickteig Makofski, "His Art Is His Joy: It Just Didn't Fit My Plan," *The New York Times*, 2005

Mimicking those experiments done to animals to put them into a state of stress, we would do basically the same things to these [tube] circuits, and you could hear them literally shrieking! It was like they were alive. . . . The same conditions that would produce breakdowns and malfunctions in machines, made for some wonderful music. The circuits would have a "nervous breakdown."

—Bebe Barron, music/sound creator of the 1956 science-fiction movie *Forbidden Planet*[1]

When it comes to art, we are a nation of extremists. American writing, painting and music have always swung between the minimal and the maximal, the Apollonian and the Dionysian. . . . We like tiny, well-made stuff and also great sprawling messes; art that is full of feeling and also art that aspires to a kind of icy perfection. The result is a culture that has given us both Audubon and Bierstadt; Dickinson and Whitman; Hemingway and Dreiser; Philip Glass and Leonard Bernstein.

–Charles McGrath, "The Souped-Up,
Knock-Out, Total Fiction Experience"[2]

Heaven is not a place but a state of consciousness.

–Abdu'l-Baha, *Divine Philosophy*

Chapter 3: Gilgamesh, God, and the Language of Angels

This has always puzzled me about these mad scientists [in movies] who are intent on playing God or unlocking the secrets of the universe or getting free cable or something. First of all, why would they bother with a wife; second, why would she bother with him, and third, why would she stay with him when it becomes apparent that he is more interested in stimulating a monkey brain than her?

–Jon Wagner, "MonsterHunter Review of *Donovan's Brain*"[3]

Language-dominated cognition among human beings enables a "mythic culture" whose primary function is to pass collective knowledge about survival through a vast mythic heritage, complete with oral lore, totemic art, mimetic song, dance, and ritual. . . . Myths, while enabling survival, also serve as carriers of important information about real events and observations.

–Abigail A. Baird, "Sifting Myths for
Truths About Our World" [4]

When we die, it's possible that it may be rather like being in a dream from which we can't wake up. . . . We're not really going to find out until we do die, and what happens then may indeed depend on our expectations. . . . It may be that the afterlife is heavily conditioned by our expectations and beliefs, just as our dreams are.

–Rupert Sheldrake, in David Jay Brown's *Conversations at the Edge of the Apocalypse*

[A myth is] an aesthetic device for bringing the imaginary but powerful world of preternatural forces into a manageable collaboration with the objective, experienced facts of life in such a way as to excite a sense of reality amenable to both the unconscious passions and the conscious mind. To this end, humans have been equipped with an ability, which lies somewhere between curiosity and compulsion, to discern meaning in our experience.

–Abigail A. Baird, "Sifting Myths for Truths About Our World"[5]

If personality exists after what we call death, it is reasonable to conclude that those who leave the Earth would like to communicate with those they have left here. I am inclined to believe that our personality hereafter will be able to affect matter. If this reasoning be correct, then, if we can evolve an instrument so delicate as to be affected by our personality as it survives in the next life, such an instrument, when made available, ought to record something.

–Thomas Edison, *Scientific American*, October 1920

I'm pretty sure that people gain a selective advantage from belief in things they can't prove. Those who are occasionally consumed by false beliefs do better in life than those who insist on evidence before they believe and act.

–Randolph M. Nesse in John Brockman's *What We Believe But Cannot Prove*

If you wake somebody up from a deep, stage-four sleep, they will usually have nothing to report about the experience. And yet the brain is not any less active than in waking. Where does consciousness disappear to during sleep?

—Giulio Tononi, "No Rest for the Snooze Guru,"
Discover, August 2005

A group of children who were tested on death comprehension reflected on what it might be like to be dead with references to "sleeping," feeling "peaceful," or simply "being very dizzy. . . ." Younger children are more likely to attribute *mental states* to a dead agent than are older children. . . . It seems that the default cognitive stance is reasoning that human minds are immortal. . . .

—Jesse Bering in John Brockman's
What We Believe But Cannot Prove

If you stay in Larissa, you will find peace. You will find a wonderful woman, and you will have sons and daughters, who will have children. And they'll all love you and remember your name. But when your children are dead, and their children after them, your name will be forgotten. . . If you go to Troy, glory will be yours. They will write stories about your victories in thousands of years! And the world will remember your name.

—Thetis, the mother of Achilles, in *Troy: The Movie*, 2004

All mythology speaks of another plane that exists alongside our own world, and that in some sense supports it.

—Karen Armstrong, *A Short History of Myth*

Chapter 4: The Matrix, Quantum Resurrection, and the Quest for Transcendence

Researchers using MRI brain scans found that some seriously brain-damaged—but not comatose—patients responded to a loved one's voice in patterns of neural activity similar to those of healthy subjects. Such scans may play a key role in future decisions over care for the unconscious.

—David Bjerklie, *Time* magazine, February 21, 2005

The sailing ship, the distant view, the lonely walks in autumn, the relative silence, it is paradise.

—Albert Einstein[6]

By 2100, the vast majority of "people" will be immortal computers running brain simulations. Simulated brains are potentially immortal, just as all computer data is. The population of such "uploads" should expand very rapidly, allowing huge increases in both economic growth rates and inequality.

—Robin Hanson, "Fourteen Wild Ideas:
Five of Which Are True!"[7]

Some say the tunnel [in near-death experiences] is a symbolic representation of the gateway to another world. But then why always a tunnel and not, say, a gate, doorway, or even the great River Styx? Why the light at the end of the tunnel? And why always above the body, not below it?

—Susan Blackmore, "Near Death Experiences:
In or Out of the Body," *Skeptical Inquirer*

There is nothing that you do, there is no thought that you have, there is no awareness, there is no lack of awareness, there is nothing that marks your daily existence that doesn't have a neural code. The greatest challenge for us is to figure out how to design the study that will reveal these codes.

—Clinton Kilts, in "Mysteries of the Mind"[8]

What we think of as "mind" is only a sort of jumped-up gland, piggybacking on the reptilian brainstem and the older, mammalian mind, but our culture tricks us into recognizing it as all of consciousness. The mammalian spreads continent-wide beneath it, mute and muscular, attending its ancient agenda.

—William Gibson, *Pattern Recognition*

A honeybee's brain may fit on the head of a match, but a research team says that the bee's working memory is almost as effective as that of a pigeon or a monkey.

—Susan Milius, "Little Brains That Could: Bees Show Big-Time Working Memory," *Science News*, April 2, 2005

Shocking and uncouth black beings with smooth, oily, whale-like surfaces, unpleasant horns that curved inward toward each other, bat-wings whose beating made no sound, ugly prehensile paws, and barbed tails that lashed needlessly and disquietingly. And worst of all, they never spoke or laughed, and never smiled because they had no faces at all to smile with, but only a suggestive blankness where a face ought to be. All they ever did was clutch and fly and tickle; that was the way of night-gaunts.

—H. P. Lovecraft, "The Dream-Quest of Unknown Kadath"

Here in the twenty-first century we like to view ourselves as ambulatory brains, plugged into meat-puppets that lug our precious grey matter from place to place. We tend to think of that grey matter as transcendently complex, and we think of it as being the bit that makes us *us*.

—Cory Doctorow, "Thought Experiments:
Interview with Kurzweil"[9]

We seem to have found one area of the brain closely associated with higher-level emotions, the spindle cells, deeply embedded in the brain. There are tens of thousands of them, spanning the whole brain . . . , which is an incredibly small number. Babies don't have any, most animals don't have any, and they likely only evolved over the last million years or so. Some of the high-level emotions that are deeply human come from these.

—Ray Kurzweil, "Thought Experiments:
Interview with Kurzweil"[10]

Daniel Dennett's view is that. . . consciousness of an event is a matter of large areas of the brain being influenced by one of a set of competing sensory signals. None of this can happen, he insists, without language. He embraces the surprising conclusion that the non-verbal (animals, infants and the profoundly aphasic) are not genuinely conscious. He has convinced himself that acquiring a language rewires the brain to make it function as a serial digital computer, and that this rewiring is necessary for consciousness.

—Patricia Churchland, "Brains Wide Shut?"[11]

The morphic fields of perception and our behavior are rooted in our brains—just like magnetic fields are rooted in magnets . . .— but nevertheless stretch out beyond their surface. I think our minds are rooted in our brains during our normal waking life, and stretch out beyond their surface through fields. So in that sense, the field concept and the soul concepts are indeed related.

—Rupert Sheldrake, in David Jay Brown's *Conversations at the
Edge of the Apocalypse*

I was six years old when my parents told me that there was a small, dark, jewel inside my skull, learning to be me. Microscopic spiders had woven a fine golden web through my brain, so that the jewel's teacher could listen to the whisper of my thoughts. The jewel itself eavesdropped on my senses. . . . I thought: if hearing that makes *me* feel strange and giddy, how must it make *the jewel feel?* Exactly the same, I reasoned; it doesn't know it's the jewel, and it too wonders how the jewel must feel . . . —it too wonders whether it's the real me, or whether in fact it's only the jewel that's learning to be me.

–Greg Egan, "Learning to be Me," in *Axiomatic*, 1995

I don't think the mind is in the brain. I think in an ordinary act of vision, when we look at something, the mind extends beyond our brain. . . . So it may be true to say near-death experiences, visionary experiences, and DMT trips are all in the mind, but that doesn't mean to say they're all in the brain.

–Rupert Sheldrake, in David Jay Brown's *Conversations at the Edge of the Apocalypse*

Forty percent of the synapses in a mouse's brain change over the course of a few weeks. Still, the mouse retains earlier memories. . . . Perhaps memories migrate to new groups of brain cells or become stored in some more efficient way. How can we have a memory that lasts for years when the underlying synapses are no longer there?

–Neurobiologist Karim Nader, "Wanted: The Next Einstein"[12]

[Future computers] will understand themselves actually better than we understand ourselves. . . . How well do we understand how our own intelligence works? They'll understand how humans work very well. So, in that sense, their self-awareness will be much greater than human self-awareness is today.

–Ray Kurzweil in David Jay Brown's *Conversations at the Edge of the Apocalypse*

It is ironic that we will admit that every mental ability can be destroyed by damage to the brain, but we tend to doubt that all of our mental abilities arise from the ordinary matter that makes up our brains. We may admit that our feelings can be altered in many ways by drugs, but we are hesitant to believe that naturally occurring brain chemicals can explain why we feel the way we do. We often feel that something else is needed . . . and that "mere matter" cannot account for our beliefs, values or feelings.

–Jerome Elbert "Does the Soul Exist?"
in Paul Kurtz's *Science and Religion*

An astronomically overwhelming majority of the people who could be born never will be. You are one of the tiny minority whose number came up. Be thankful that you have a life, and forsake your vain and presumptuous desire for a second one.

–Richard Dawkins, "The Atheist," Salon.com[13]

The Multiverse idea . . . says that in the beginning, there was nothing . . . but hyperspace–unstable because of the quantum principle. . . . This means that bubbles began to form in nothing, and these bubbles began to expand rapidly, giving us the Universe. . . . The Judeo-Christian genesis takes place within the Buddhist nirvana . . . and our Multiverse percolates universes.

–Michio Kaku, "Parallel Universes, the Matrix, and
Superintelligence," KurzweilAI.net[14]

We have begun to see that almost any kind of material can serve as a computer. Human brains, which are mostly water, compute fairly well. . . . So can stick and strings.

–Kevin Kelly, "One Universal Computation," in
Spiritual Information, edited by Charles L. Harper, Jr.

[Many radio frequencies] exist simultaneously all around you. . . . However, your radio is only tuned to one frequency. In the same way, in your living room . . . there is the wave function of aliens from outer space. There is the wave function of the Roman Empire, because it never fell. . . . However, just like you can only tune into one radio channel, you can only tune into one reality channel, and that is the channel that you exist in.

—Michio Kaku, "Parallel Universes, the Matrix, and Superintelligence," KurzweilAI.net[15]

The sun turned the stained glass window of the chapel into matrices of burning jewels. . . . Daniel saw in a way he'd never seen anything before. His mind was a homunculus squatting in the middle of his skull, peering out through good but imperfect telescopes. . . .

—Neal Stephenson, *Quicksilver*

When we go to sleep and we wake up the next day, we usually wind up in the same universe. . . . You could never rule out the fact that the world could be a dream, but the fact of the matter is, the universe as it exists is a reproducible universe.

—Michio Kaku, "Parallel Universes, the Matrix, and Superintelligence," KurzweilAI.net[16]

58%   Number of physicists (including Stephen Hawking) who think multiple universes exist[17]

18%   Number of physicists (including Roger Penrose) who do not accept the existence of parallel universes

13%   Number of physicists who admit the possibility, but remain unconvinced

11%   Number of physicists who have no opinion

Suppose for a moment that children in the future will grow up with an easy and intimate virtual-reality technology and that their use of it will become focused on invention and design instead of the consumption of pre-created holo-video games. . . . Maybe these future children will play virtual musical-instrument-like things that cause simulated trees and spires and seasons and odors and ecologies to spring up. . . .

–Jaron Lanier in John Brockman's *What We Believe but Cannot Prove*

I certainly don't believe in the soul as an enduring entity. Our brains are made of neurons, and nothing else. Nerve cells are very complicated mechanical systems. You take enough of those, and you put them together, and you get a soul.

–Daniel Dennett, "The Nonbeliever,"
The New York Times Magazine, January 22, 2006

They say that dreams are only real as long as they last. Couldn't you say the same thing about life?

–Richard Linklater, *Waking Life*

A dream is an answer to a question we haven't learned how to ask.

–FBI Agent Dana Scully, "Paper Hearts,"
The X-Files TV show

Scientists have looked at [the universe] as a ragtag collection of particles and fields while failing to see what it is as a majestic whole: an enormous computer. Every physical event, everywhere, feeds information into it. And the output of the cosmic computer is nothing less than reality itself.

–Corey S. Powell, "Welcome to the Machine,"
The New York Times Book Review, April 2, 2006

How [could] the brain *not* be conscious when it has over 100 billion neurons, each connected to 10,000 others and firing at rates of 2–60 times every second? The information processed is more than 10^{29} bits/second. . . ! "That's enough information to support both perception and the perception of perception, and enough to support awareness and awareness of awareness."

–Bill and Rich Sones, quoting J. Allan Hobson, "The HooK:
Strange But True," March 30, 2006

[MIT scientist] David Cory has used a simple quantum computer to study how information flows through the subatomic world. If these devices truly match the workings of the universe, expanded versions could be used, for example, to develop a more complete theory of gravity, whose essence is still utterly mysterious.

–Corey S. Powell, "Welcome to the Machine," *The New York
Times Book Review*, April 2, 2006

The dead will be resurrected when the computer capacity of the universe is so large that the amount of capacity required to store all possible human simulations is an insignificant fraction of the entire capacity. Since the information storage capacity diverges to plus infinity. . . , resurrection will occur between 10 raised to the power of -10^{10} seconds and 10 raised to the power of -10^{123} seconds before the Omega Points is reached.

–Frank Tipler, *The Physics of Immortality*

Consciousness is not neurons firing–consciousness is a transcendent emergent phenomenon that depends on the firing of neurons. The gears of a watch rotate and keep time, but the turning of the gears is not time. The question is, Is neuroimaging a picture of the experience of consciousness or is it

a picture of a mechanism associated with the experience? . . .
If a higher being told us how consciousness works, could we
understand the explanation?
    —Daniel Carr in Melanie Thernstrom's "My Pain, My Brain,"
                    The New York Times Magazine, May 14, 2006

When he . . . had cut open live dogs during the Plague Year,
Daniel had looked into their straining brown eyes and tried to
fathom what was going on in their minds. He'd decided, in the
end, that *nothing* was, that dogs had no conscious minds, no
thought of past or future, living purely for the moment, and that
this made it worse for them. Because they could neither look forward to the end of the pain, nor remember times when they had
chased rabbits across meadows.
                    —Neal Stephenson, *Quicksilver*

If the universal resurrection is accomplished by reassembling the
original atoms which made up the dead, would it not be logically impossible for God to resurrect cannibals? Every one of
their atoms belongs to someone else!
                    —Frank Tipler, *The Physics of Immorality*

I saw Eternity the other night,
Like a great ring of pure and endless light,
All calm, as it was bright;
And round beneath it, Time in hours, days, years,
Driv'n by the spheres
Like a vast shadow mov'd; in which the world
And all her train were hurl'd. . . .
Yet some, who all this while did weep and sing,
And sing, and weep, soar'd up into the ring;
But most would use no wing.

O fools (said I) thus to prefer dark night
Before true light,
To live in grots and caves, and hate the day|
Because it shews the way,
The way, which from this dead and dark abode
Leads up to God,
A way where you might tread the sun, and be
More bright than he.
But as I did their madness so discuss
One whisper'd thus,
"This ring the Bridegroom did for none provide,
But for his bride."

                 —Henry Vaughan (1621–1695)[18]

Theoretical physicists working in the rarefied field of loop quantum gravity have developed a way to describe elementary particles as mere tangles in space. . . . Everything in the universe emerges from a simple network of relationships, with no fundamental building blocks at all. They are not even relationships between objects as such, just an abstract graph of connections. . . . When a network is tied in a braid, it forms something like a particle. . . . Space itself becomes just a web of information.

          —David Castelvecchi, "Out of the void," *New Scientist*,
                 August 12, 2006 (and also the associated
                       unsigned editorial piece in this issue)

Chapter 5: Jesus and the Future of Mind-Altering Drugs

People who are smoking pot aren't eager to turn themselves in. And their friends aren't eager to turn them in either, in most cases. The whores don't want to turn in the Johns; the Johns don't want to turn in the whores. The gamblers don't want to turn in the bettors; the bettors don't want to turn in the bookies, and so on. . . . The only way you can wage a war against sin—which is what these victimless crime laws are all about—is by spying on every body more and more. . . . Even Franz Kafka couldn't imagine a society so crazy that you have to give urine samples before you can hold a job.

—Robert Anton Wilson in Richard Metzger's
Disinformation: The Interviews

If people take Viagra to get hard, caffeine to wake up, CX717 to do better on SAT tests, and aspirin to take away pain, why can't they take DMT to transcend space and time?

—Anonymous

At the beginning of the 20th century, there were only two psychedelic compounds known to Western science: cannabis and mescaline. A little over 50 years later—with LSD, psilocybin, psilocin, TMA, several compounds based on DMT and various other isomers—the number was up to almost 20. By 2000, there were well over 200. So you see, the growth is exponential.

—Alexander Shulgin, "Dr. Ecstasy"[19]

Many people die from rock climbing. Some people die from taking psychoactive drugs. We outlaw these drugs. Why don't we outlaw the rock climbing?

—Anonymous

Many patients with difficult to treat conditions use cannabis to relieve their symptoms, but in most parts of the world that makes them criminals. Otherwise law-abiding citizens dislike having to get their treatments from drug dealers.

—Clare Wilson, "Miracle Weed," *New Scientist*, February 5, 2005

In 2002 there were 1,538,813 total drug arrests, according to the FBI Uniform Crime Reports, an astounding number, particularly when you think about all those that didn't get caught. A full 80% of these were for mere possession of a controlled substance, and 50% of those, 613,986 people, were for nothing more than possession of marijuana.

–Charles Shaw, "The Vice Lords of the Replacement Economies," guerrillanews.com, June 2005

If cannabis were unknown, and bioprospectors were suddenly to find it in some remote mountain crevice, its discovery would no doubt be hailed as a medical breakthrough. Scientists would praise its potential for treating everything from pain to cancer, and marvel at its rich pharmacopoeia. . . .

–"Reefer Madness," *The Economist*, April 27, 2006

Nearly all kinds of patients with neurotic and psychosomatic disease can be helped [by hallucinogenic-assisted psychotherapy], as shown by the 300 to 400 studies from the 1950s to 1960s.

–Professor Torsten Passie, Hanover Medical School, "Psychedelic Healing," *New Scientist*, 190(2547): 50, April 15, 2006

I estimate a 99 percent probability that we can [while in altered states of consciousness] communicate with intelligences seemingly not our own. I've done it. I still remain unsure, however, if the Higher Intelligences contacted dwell in outer space, in other dimensions, or in circuits of my own brain not identifiable as "me" or "my ego." I'd love to live in an open democratic society where research on this becomes legal and widely published.

–Robert Anton Wilson, in David Jay Brown's *Conversations at the Edge of the Apocalypse*

From my rotting body, flowers shall grow and I am in them and that is eternity.

—Edvard Munch (1863–1944), Norwegian painter

Chapter 6: Clockwork Butterflies and Eternity

If we look for insights into human nature to guide the future of religion, we shall find more such insights in the novels of Dostoevsky than in the journals of cognitive science. Literature is the great storehouse of human experience, linking together different cultures and different centuries. . . . Literature enables us to share the passions of Greek and Trojan warriors . . . and of Hebrew prophets and kings. . . . Literature will remain as the way we embalm our thoughts and feelings for transmission to our descendants. Literature survives when the civilizations that gave birth to it collapse and die.

—Freeman Dyson, "Complementarity," in *Spiritual Information*, edited by Charles L. Harper, Jr.

Within a few decades, we should be able to create robots as smart as mice. . . . However, when machines start to become as dangerous as monkeys, I think we should put a chip in their brain, to shut them off when they start to have murderous thoughts. By the time you have monkey intelligence, you begin to have self-awareness, and . . . an agenda created by a monkey for its own purposes.

—Michio Kaku, "Parallel Universes, the Matrix, and Superintelligence," KurzweilAI.net[20]

Once upon a time, I, Chuang Tzu, dreamt I was a butterfly. . . . Soon I awaked, and there I was, myself again. Now I do not know whether I was then a man dreaming I was a butterfly, or whether

I am now a butterfly, dreaming I am a man. Between a man and a butterfly there is necessarily a distinction. The transition is called the transformation of material things.

–Chuang Tzu (c. 369–c. 286 BC), the most significant of China's early interpreters of Taoism

Open your eyes; listen, listen. That is what the novelists say. But they don't tell you what you will see and hear. All they can tell you is what they have seen and heard, in their time in this world, a third of it spent in sleep and dreaming, another third of it spent in telling lies. . . .

The artist deals with what cannot be said in words. The artist whose medium is fiction does this in words. The novelist says in words what cannot be said in words.

–Ursula K. Le Guin, Introduction to *The Left Hand of Darkness*

Chapter 7: Evolution, Ice Cream, and the Goddess of Chopped Liver

The most highly developed parts of the human frontal cortex that deal with decisions and social interactions are right next to the parts that control taste and smell and movements of the mouth, tongue and gut. There is a reason we kiss potential mates—it's the most primitive way we know to check something out.

–Helen Phillips, "Life's Greatest Invention," *New Scientist*[21]

I see evolution as a spiritual process, because it moves toward greater and greater qualities that we associate with God. . . . We see that evolution creates entities that become. . . more creative and intelligent over time.

–Ray Kurzweil in David Jay Brown's *Conversations at the Edge of the Apocalypse*

Many parasites, such as the small liver fluke, have also mastered the art of manipulating their host's behavior. Ants whose brains are infected with a juvenile fluke feel compelled to climb to the tops of grass blades, where they are more likely to be eaten by the fluke's ultimate host, a sheep.

—Anna Gosline, "Life's Greatest Invention," *New Scientist*[22]

The bacteria in our guts outnumber our own cells by nearly 10 to 1—and nearly two-thirds of them are new to science.

—"Gut Bug Bonanza," *New Scientist*, April 23, 2005

In 1995, Americans consume more than 5 billion pounds of deli meats a year, including about 800 million pounds of bologna.

—"Food Facts and Trivia: Delicatessen"[23]

Box jellyfish, or cubozoans, are fantastic creatures with 24 eyes, four parallel brains, and 60 arseholes.

—Dan Nilsson, "Multi-eyed Jellyfish
Casts New Light on Darwin's Puzzle"[24]

The distribution of species on islands and continents throughout the world is exactly what you'd expect if evolution was a fact. The distribution of fossils in space and in time are exactly what you would expect if evolution were a fact. There are millions of facts all pointing in the same direction and no facts pointing in the wrong direction.

British scientist J. B. S. Haldane, when asked what would constitute evidence against evolution, famously said, "Fossil rabbits in the Precambrian." They've never been found. Evolution is a fact.

—Richard Dawkins, "The Atheist," Salon.com[25]

One could also claim that life is so complex that no conceivable intelligence could design it, so it must have evolved from simpler material. In fact, I'm making this my new religion and will have blind faith in it for the rest of life.

—Chuck Gaydos, personal communication, 2005

Resistance to evolution comes from religion. And from bad religion. You won't find any opposition to the idea of evolution among sophisticated, educated theologians. It comes from an exceedingly retarded, primitive version of religion, which unfortunately is at present undergoing an epidemic in the United States. Not in Europe, not in Britain, but in the United States.

—Richard Dawkins, "The Atheist," Salon.com[26]

So if God is everywhere, then why is God so hard to perceive? One could imagine a God who would be more like a Chairman Mao or a Comrade Stalin. This God would have designed a universe with photographs of himself hung everywhere in nature. . . . Of course, science is yet to find an unequivocal "made by God" label attached to nature.

—William Grassie, "The Problem with Intelligent Design," Metanexus.net

This new attack claims that because there are gaps in evolution, they therefore must be filled by a divine intelligent designer. How many times do we have to rerun the Scopes "monkey trial"? There are gaps in science everywhere. Are we to fill them all with divinity? There were gaps in Newton's universe. They were ultimately filled by Einstein's revisions. There are gaps in Einstein's universe, great chasms between it and quantum theory. Perhaps they are filled by God. Perhaps not. But it is certainly not science to merely declare it so.

—Charles Krauthammer, "Let's Have No More Monkey Trials," *Time*, August 8, 2005

You can't statistically explain improbable things like living creatures by saying that they must have been designed because
you're still left to explain the designer, who must be, if anything,
an even more statistically improbable and elegant thing.

> —Richard Dawkins, "The Atheist," Salon.com[27]

Who designed the cosmic designer [responsible for creating life]?
If the answer, most probably, is that no one designed the
designer, then one can legitimately ask the follow-up question,
"If a designer is not needed to design the designer, why is a
designer needed to design a butterfly?"

> —Michael Abraham, "Darwin and Design," *New Scientist*
> Letters section, July 30, 2005

Why would the designer give us a [biochemical] pathway for
making vitamin C, but then destroy it by disabling one of its
enzymes?

> —Jerry Coyne, "The Case Against Intelligent Design,"
> Edge.Org and *The New Republic*, August 2005

It seems to me that if Intelligent Design advocates accept the
time periods associated with evolution, rather than the pure
Judaeo-Christian seven-day creation, then they have created a
new question that they must answer: why did the creation of
humans take so long?

> —Nicholas Adams, "Darwin and Design," *New Scientist*
> Letters section, July 30, 2005

54% of American adults did not believe humans had developed
from an earlier species.

> —Harris poll, 2005[28]

Artificial selection used by breeders had wrought immense
changes in plants and animals. . . . Starting with wild cabbage,

breeders produced . . . broccoli, cauliflower, and Brussels sprouts. Artificial selection is nearly identical to natural selection, except that humans rather than the environment determine which variants leave offspring. And if artificial selection can produce such a diversity . . . in a thousand-odd years, natural selection could obviously do much more over millions of years.

–Jerry Coyne, "The Case Against Intelligent Design,"
Edge.Org and *The New Republic*, August 2005

"I stand in my synagogue and pray to God and have an intense relationship with God, and yet I don't believe in God." [Philosopher Lipton] compared his religious experience with that of someone who gets pleasure and meaning from a novel even though he knows it is not literally true.

–John Horgan, quoting Cambridge philosopher Peter Lipton,
Scientific American, September 2005

As a Jew, I have no problem whatsoever believing in intelligent design. . . . I chose to believe that evolution was packaged with the original matter that resulted in the big bang. The "designer" can no longer intervene because he is held by the laws of the universe he willed.

–Epidemiologist Marysia Meylan, *The New York Times*,
September 4, 2005

Five months after conception, human fetuses grow a thin coat of hair, called lanugo, all over their bodies. It does not seem useful—after all, it is a comfortable 98.6 degrees in utero—and the hair is usually shed shortly before birth. The feature makes sense only as an evolutionary remnant of our primate ancestry; fetal apes also grow such a coat, but they do not shed it.

–Jerry Coyne, "The Case Against Intelligent Design,"
Edge.Org and *The New Republic*, August 2005

If science had never observed the formation of emeralds in a natural setting—but *theorized* that emeralds were created by natural forces over millions of year—would proponents of Intelligent Design posit the supernatural creation of emeralds? After all, emeralds ($Be_3Al_2Si_6O_{18}$) embody a gorgeous and complex crystalline form, and contain the rare element beryllium with a trace of chromium or vanadium. How could it be just chance formation the led to such complexity and beauty?

—Cliff Pickover

If creationists accept that an eye or heart can form spontaneously through physical laws from a fertilized egg, why do they find it impossible that the eye or heart can form spontaneously through physical laws in a process of evolution? Both processes generate amazing complexity and structures.

—Cliff Pickover

Intelligent Design is bad theology because it turns God into a mere garage tinkerer, a fumbling watchmaker, a Dr. Frankenstein cobbling together biochemical parts from the primordial soup into complex organisms. Such a God cannot be the omniscient and omnipotent God of Abraham; indeed, the Intelligent Design God would have the same skill sets as an advanced extraterrestrial intelligence capable of genetic engineering and other feats. If God is the maker of all things visible and invisible in heaven and earth, God must be above such restraints; that is, above the laws of nature and contingencies of chance.

—Michael Shermer, *The New York Times Book Review*,
Letters section, February 5, 2006

Intelligent design seeks to answer the question, "Does a system (the universe, or a biological system, or a radio signal, or an oddly shaped rock) show signs of being designed, rather than merely being ordered?" It attempts to . . . evaluate potential answers to the question through . . . statistical and probability theory. But, when a biologist applies the principles . . . to an analysis of a highly ordered biological system and dares to ask the question, "Is this sequence . . . the product of chance, or is this sequence intelligently designed?" then the evolution-only dogmatists decry the very methods that they find no quarrel with in other scientific disciplines.

–David Medici, "Letters: Intelligent Design Is Most Definitely Scientific," February 5, 2006, Sheboygan-Press.com

Ninty-nine percent of all living creatures die by being eaten alive.

–Mary-Ann Tirone Smith, *She's Not There*

If one was able to replay the whole evolution of animals, starting at the bottom of the Cambrian (and. . . . moving one of the individual animals two feet to its left), there is no guarantee–indeed, no likelihood–that the result would be the same. There might be no conquest of the land, no emergence of mammals, and certainly no human beings.

–John Maynard Smith, "Taking a Chance on Evolution," *The New York Review of Books*

One of the favorite phrases of Intelligent Design is "irreducible complexity," and surely, that would apply to G.O.D., implying that he must have been designed in turn by another intelligent designer, who, of course, must have had his own intelligent designer, and so on, *ad infinitum.*

–John G. Bentley, *Skeptical Inquirer*, May/June 2006

In Iceland, Denmark, Sweden, and France, 80% or more of adults accepted the concept of evolution. [On the other hand], one in three American adults firmly rejects the concept of evolution.

–Jon Miller, Eugenie Scott, Shinji Okamoto, "Public Acceptance of Evolution," *Science*, 313(5788): 765, August 11, 2006

Chapter 8: The Whispers of History

In the last 3,421 years of recorded history only 268 have seen no war.
> —Will and Arial Durant, *The Lessons of History*

If we are getting restless under the taxonomy of a monocotyledonous wage doctrine and a cryptogamic theory of interest, with involute, loculicidal, tomentous and moniliform variants, what is the cytoplasm, centrosome or karyokinetic process to which we may turn, and in which we may find surcease from the metaphysics of normality and controlling principles?
> —Thorstein Veblen, "Why Is Economics Not an Evolutionary Science?" *The Quarterly Journal of Economics*, Volume 12, 1898

Archeology suggests that something very special began to happen to our species around 40,000 years ago [exemplified by carvings, figurines, grave goods, ornamentation, and musical instruments such as bone flutes]. . . . If not language itself, perhaps the Great Leap Forward coincided with the sudden discovery of what we might call a new software technique: maybe a new trick of grammar, such as a conditional clause, which, at a stroke, would have enabled "what if" imagination to flower.
> —Richard Dawkins, *The Ancestor's Tale*

Ethically we have not moved beyond Jesus Christ, artistically we have not moved beyond Titian, and politically we have not moved beyond Jefferson. Scientifically, however, we are moving forward all the time.
> —Bryan Appleyard, "People in Glass Houses," *New Scientist*, October 8, 2005

This universal character is described as a vast fine-meshed net drawn through the cosmos so that everything known in heaven and earth was trapped in one of its myriad cells. All that was needed to identify a particular thing was to give its location in the table, which could be expressed as a series of numbers.

—Neal Stephenson, *Quicksilver*

Previously, alphabetic print had exploded Western culture into millions of hard-edged shards of individualistic shrapnel. Both reading and writing are, in most cases, solitary endeavors. Television abruptly reversed the process, and the centripetal implosion not only pulled together individual families but also began to enmesh the entire human community into what McLuhan called "one vast electronic global village." Television was so startlingly original that many other adjustments in perception were necessary for the brain to make sense of it.

—Leonard Shlain, *The Alphabet Versus the Goddess*

Every year . . . some new invention, method, or situation compels a fresh adjustment of behavior and ideas. Furthermore, an element of chance, perhaps of freedom, seems to enter into the conduct of metals and men. We are no longer confident that atoms, much less organisms, will respond in the future as we think they have responded in the past. The electrons, like Cowper's God, move in mysterious ways their wonders to perform, and some quirk of character or circumstance may upset national equations, as when Alexander drank himself to death and let his new empire fall apart (323 B.C.), or as when Frederick the Great was saved from disaster by the accession of a Czar infatuated with Prussian ways (1762).

—Will Durant, *The Lessons of History*

The book should be a ball of light in one's hand.
Lucky Numbers: 48, 11, 3, 44, 20, 4
                              —Message in Cliff Pickover's fortune cookie,
                         Discovered April 21, 2006, Empire Hunan Restaurant,
                                                    Yorktown Heights, New York

Conclusion

Oxford philosopher J. L. Austin noted that while a double negative amounts to a positive, never does a double positive amount to a negative.
                              —James Ryerson, "Sidewalk Socrates"[29]

My coming to New York had been a mistake; for whereas I had looked for poignant wonder and inspiration in the teeming labyrinths of ancient streets that twist endlessly from forgotten courts and squares and waterfronts to courts and squares and waterfronts equally forgotten, and in the Cyclopean modern towers and pinnacles that rise blackly Babylonian under waning moons, I had found instead only a sense of horror and oppression which threatened to master, paralyze, and annihilate me.
                                          —H.P. Lovecraft, "He"

The [conscious] simulation's internal relationship would be the same if the program were running correctly on any of an endless variety of possible computers, slowly, quickly, intermittently, or even backwards and forwards in time, with the data stored as charges on chips, marks on a tape, or pulses in a delay line, with the simulation's numbers represented in binary, decimal, or

Roman numerals, compactly or spread widely across the machine. There is no limit, in principle, on how indirect the relationship between simulation and simulated can be.

—Hans Moravec, *Robot*

Scientific fundamentalism is the belief that the world is accessible to and ultimately controllable by human reason. This is a profoundly unscientific idea. It is neither provable nor refutable. Obviously it is a leap of faith to insist that human reason is capable of fully understanding the world. We seem to have some access to its workings, but it would be wildly premature to believe that the human brain is capable of comprehending all reality.

—Bryan Appleyard, "People in Glass Houses," *New Scientist*, October 8, 2005

Girls are going to be left behind [in the future]. When we talk about people who play video games a lot, we're talking about boys. And 15 years from now, there's going to be lots of jobs in the new economy where we're going to be saying, Why are all these men getting these jobs? . . . Those are going to be the men who, as boys, played lots and lots of video games.

—Caitlin Flanagan, "Around the Corner," Interview in *Time*, March 20, 2006

Once the requisite uploading technology becomes available a few decades hence, you could make a perfect-enough copy of me. . . . The copy doesn't have to match the quantum state of my every neuron, either. . . . There are quite a few changes that each of us undergo from day to day—we don't examine the assumption that we are the same person closely.

—Ray Kurzweil, "Thought Experiments: Interview with Kurzweil"

There is no reason why the most fundamental aspects of the laws of nature should be within the grasp of human minds, which evolved for quite different purposes, nor why those laws should have testable consequences at the moderate energies and temperatures that necessarily characterize life-supporting planetary environments. . . . As we probe deeper into the intertwined logical structures that underwrite the nature of reality, we can expect to find more deep results which limit what can be known. Ultimately, we may even find that their totality characterizes the universe more precisely than the catalogue of those things that we can know.

–John Barrow, *Boundaries and Barriers: On the Limits of Scientific Knowledge*

The work of scientists and scholars is mined by the news media and promulgated to the general public, often emphasizing "factoids" that have the power to amaze. . . . To give an example, giant pandas are prominent items of popular culture; parasitic worms, though of greater practical importance, are not.

–"Pop Culture," from *Wikipedia, the Free Encyclopedia*[30]

Perhaps a simple differential equation [exists] whose solution implies our whole physical universe and everything in it. . . . As our brains and bodies cease to function . . . it takes greater and greater . . . coincidences to explain continuing consciousness by their operation. We lose our ties to physical reality, but, in the space of all possible worlds, that cannot be the end. Our consciousness continues to exist in some of those, and we will always find ourselves in worlds where we exist and never in one where we don't.

–Hans Moravec, *Robot*

NOTES
❦

Chameleon tongues are extremely fast and long. They can be anywhere from one to 1.5 times the body length of the owner and can rocket in and out with blinding speed. A 5.5-inch tongue reaches full extension in $\frac{1}{16}$ of a second, which is fast enough to snatch a fly in midair.

—"The Reptipage," Reptilis.net

I've compiled a list of reference materials and notes that identify much of the material I used to research this book. I also include Internet Web sites in addition to books and journals. As many readers are aware, Internet Web sites come and go. Sometimes they change addresses or completely disappear. The Web site addresses listed in this book provided valuable background information when this book was written. You can, of course, find numerous other Web sites relating to the curiosities discussed in this book by using standard Web search tools.

If I have overlooked an interesting cultural curiosity, chameleon person, reference, or factoid—which you feel has never been fully appreciated—please let me know. Just visit my Web site, www.pickover.com, and send me an e-mail explaining the idea and how you feel it influences our perception of reality or has affected our popular culture.

Epigraph Page
1. "*Dream Dictionary* Entry for Chameleon,"
www.hyperditionary.com/dream.

Introduction
1. Siegel, Marvin. *The Last Word: The New York Times Book of Obituaries and Farewells: A Celebration of Unusual Lives.* (New York: William Morrow & Company, 1997).

2. "*Publishers Weekly* List of Best-selling Hardcover Books for 1900–1995," www.caderbooks.com/bestintro.html.

3. You can turn the equation upside down.

4. Pink, Daniel. "Revenge of the Right Brain," *Wired* 13(2): 70–72, February 2005.

5. Betts, Kate. "Yoga's Growing Reach," *Time* 165(5): 74, January 31, 2005.

6. Hodgman, John. "The Haunting," *The New York Times Magazine*, Section 6, 22–27, July 23, 2006.

7. Stanley, Alessandra. "Once Again, Having Its 7 Minute Flame," *The New York Times*, Saturday, Section E1, "The Arts," December 25, 2004. (Discusses the WPIX Yule log.)

8. According to the US Census Bureau (www.census.gov/cgi-bin/ipc/pcwe), about 360,000 people are born each day in the world. The actual number of people *conceived* each day is roughly double this number because approximately 50 percent of all fertilized eggs die and are lost (aborted) spontaneously, usually without the woman knowing she is pregnant.

9. Levitt, Steven and Stephen Dubner. *Freakonomics: A Rogue Economist Explores the Hidden Side of Everything.* (New York: William Morrow, 2005), 5–6.

10. Ibid.

11. Mitchell, Stephen. *Gilgamesh: A New English Version.* (New York: Free Press, 2004), 27.

Chapter 1. Truman Capote and *The Brain from Planet Arous*

1. "American Masters—Truman Capote/PBS," www.pbs.org/wnet/americanmasters/database/capote_t.html.

2. Krebs, Albin. "Obituary: Truman Capote Is Dead at 59; Novelist of Style and Clarity," *The New York Times*, www.nytimes.com/learning/general/onthisday/bday/0930.html.

3. Ibid.

4. Ibid.

5. "American Masters" Truman Capote/PBS,"
www.pbs.org/wnet/americanmasters/database/capote_t.html.

6. Capote, Truman. *In Cold Blood* (New York: Vintage, reprint edition, 1994), 5.

7. See for example, Stephanie Chin. "What the Academy Wants: This Year's Oscar Frontrunners," *The Tufts Observer*, March 03, 2006, www.tuftsobserver.org/arts/20060303/what_the_ academy_wants_th.html. (Other reviewers have made similar observations regarding Capote becoming immortal through *In Cold Blood.*)

8. Conway, Jeffery, Lynn Crosbie, David Trinidad. *Phoebe 2002: An Essay in Verse* (New York: Turtle Point Press, 2003). See also David Trinidad. "The Marriage of Heaven and Hell," www.lapetitezine.org/DavidTrinidad.htm.

9. Haskell, Molly. "Unmourned Losses, Unsettled Claims," *The New York Times*, June 12, 1988, Section 7, p. 1, (Book Review Desk). See also, Gerald Clarke. *Capote: A Biography.* (New York: Simon & Schuster, 1988). http://partners.nytimes.com/books/97/12/28/home/ capote-biography.html.

10. Ibid.

11. Mendelsohn, Daniel. "The Truman Show," *The New York Times Book Review*, Section 7, 15–17, December 5, 2004.

12. Ibid.

13. Capote, Truman. *The Complete Stories of Truman Capote* (Introduction by Reynolds Price) (New York: Random House, 2004).

14. Scott, Robert. "The Work Habits of Highly Successful Writers," *Writer's Digest*, 85(5): 33–35, May 2005.

15. Ibid.

16. Martin, Doug. "The Blackwing 602–the Final Chapter," June, 2004, www.pencilpages.com/articles/blackwing.htm.

17. Ibid.

18. Ibid.

19. Ludwig, Arnold. *The Price of Greatness: Resolving the Creativity and Madness Controversy* (New York: Guilford Press, 1995), 45. Also see Pickover, Clifford, *Strange Brains and Genius: The Secret Lives of Eccentric Scientists and Madmen* (New York: Quill, 1999); Pickover, Clifford, *Sex, Drugs Einstein and Elves* (Petaluma, California: Smart Publications, 2005).

20. Jacobs, A. J. *The Know-It-All: One Man's Humble Quest to Become the Smartest Person in the World* (New York: Simon and Schuster, 2004), 191.

21. The results of various finger length studies can be found on the World Wide Web. The findings in this paragraph come from: "Finger Length 'Key to Aggression,'" *BBC News*, March 4, 2005, http://news.bbc.co.uk/2/hi/health/4314209.stm; "Hidden significance of a man's ring finger," *Research Intelligence*, Issue 2, September 1999, www.liv.ac.uk/researchintelligence/issue2/finger.html; Martin S., Manning J., and Dowrick C., "Fluctuating Asymmetry, Relative Digit Length and Depression in Men," *Evolution and Human Behaviour*, 20 (1999): 203–214; Drew Voight, "Estrogen and Testosterone Battle," www.4-men.org/testosterone/levels-of-testosterone.html.

22. Pickover, Clifford. *Strange Brains and Genius: The Secret Lives of Eccentric Scientists and Madmen.* (New York: Quill, 1999). This book also gives references to other books that discuss creative geniuses.

23. "List of famous gay, lesbian or bisexual people," Wikipedia Encyclopedia, http://en.wikipedia.org/wiki/List_of_famous_gay%2C_lesbian_or_bisexual_people.

24. Weinrich, James. "Nonreproduction, Homosexuality, Transsexualism, and Intelligence: I. A Systematic Literature search," *Homosexuality*, Spring; 3(3): 275–89, 1978. Note: my colleagues wonder if

such studies rely on the self-identification of individuals as homosexuals—and if people who recognize themselves as homosexual, and are prepared to admit it publicly, are drawn from a more highly educated section of the population.

25. Ludwig, Arnold, op. cit. *The Price of Greatness: Resolving the Creativity and Madness Controversy.* See also Pickover, Clifford, op. cit. *Strange Brains and Genius: The Secret Lives of Eccentric Scientists and Madmen.*

26. Ibid.

27. Jacobs, A. J., op. cit. *The Know-It-All: One Man's Humble Quest to Become the Smartest Person in the World,* 264.

28. Woolf, Virginia. *The Letters of Virginia Woolf, Volume 6,* (New York: Harvest Books, 1982), 481.

29. Dick, Philip K. interviewed by Arthur Byron Cover in "Vertex Interviews Philip K. Dick," *Vertex,* 1(6), February 1974, www.philipkdick.com/media_vertex.html.

30. Waldron, Ann. "Writers and Alcohol," *The Washington Post,* March 14, 1989, pp. 13–15, www.unhooked.com/sep/writers.htm.

31. Neff, Nancy. "Signs of Suicidal Tendencies Found Hidden in Dead Poets' Writings," Center for the Advancement of Health," www.hbns.org/newsrelease/signs7-24-01.cfm.

32. Ibid.

33. Kirn, Walter. "Tumultuous Lowell," *The New York Times Book Review,* Section 7, 10–11, June 26, 2005.

34. Ibid.

35. Folley, Bradley S., and Sohee Park. "Verbal Creativity and Schizotypal Personality in Relation to Prefrontal Hemispheric Laterality: A Behavioral and Near-Infrared Optical Imaging Study," *Schizophrenia Research,* available online August 24, 2005, www.sciencedirect.com/science/journal/09209964.

36. Moran, Melanie. "Odd Behavior and Creativity May go

Hand-in-Hand," *Exploration: The Online Research Journal of Vanderbilt Unviversity*, September 6, 2005, http://exploration.
vanderbilt.edu/news/news_schizotypes.htm.

37. Lehmann-Haupt, Christopher. "Books of The Times," *The
New York Times,* August 5, 1980, http://partners.nytimes.com/
books/97/12/28/home/capote-music.html.

38. Rosenthal, Elisabeth. "For Fruit Flies, Gene Shift Tilts Sex Orientation," *The New York Times*, CLIV(53, 234): A1, June 3, 2005.

39. Wade, Nicholas. "Researchers Say Intelligence and Disease
May be Linked in Ashkenazic Genes," *The New York Times*,
CLIV(53, 234): A21, June 3, 2005.

Chapter 2. John Cage and The Zen of Music

1. Glenn Gould published this quotation in an article on the future of
music recording in *High Fidelity*, April 1966. I learned about the Gould
quote from MIT Professor Tod Machover, at "The Brain Opera,"
http://brainop.media.mit.edu/libretto/todarticle.html.

2. This quotation comes from Tod Machover, at "The Brain
Opera and Active Music,"
http://brainop.media.mit.edu/Archive/ars-Electronica.html.
(Machover suggests that Cage "might have said" this. Perhaps
Machover implies that this statement provides an excellent indi-
cation of Cage's philosophy.)

3. Krukowski, Damon, Art Lange, and Joan La Barbara. "John
Cage–Europera 5," Mode Records, www.mode.com/
catalog/036cage.html.

4. Ibid.

5. Ibid.

6. John Cage. "An Autobiographical Statement." First appeared in
print in the *Southwest Review*, 1991. "An Autobiographical Statement" was written for the Inamori Foundation and delivered in
Kyoto as a commemorative lecture. John Cage also delivered "An
Autobiographical Statement" at Southern Methodist University in

1990. This statement is reprinted on the Web at New Albion Records, www.newalbion.com/artists/cagej/autobiog.html.

7. Cage, John. "Story 60," in *Silence* (Middletown, Connecticut: Wesleyan University Press, 1961). Cage wrote many paragraph-long stories, anecdotes, thoughts, and jokes, which he would read aloud. The transcript of "Story 60" (on mycology), from his book *Silence*, is available at www.lcdf.org/indeterminacy/s.cgi?60.

8. Jacobs, A. J. *The Know-It-All: One Man's Humble Quest to Become the Smartest Person in the World.* (New York: Simon and Schuster, 2004).

9. Ibid.

10. Holdrege, Craig "Genes and Life: The Need for Qualitative Understanding," *Context* #1 (Spring, 1999, pp. 11–15), reprinted at The Nature Institute site, http://natureinstitute.org/pub/ic/ic1/genes.htm.

The Goethe quote was translated by Holdrege and it appeared in Schad, W. "Stauphaenomene am menschlichen Knochenbau," in *Goetheanistische Naturwissenschaft*, Bd. 4 Anthropologie, pp. 9–29, edited by W. Schad (Stuttgart: Verlag Freies Geistesleben, 1985).

11. I discuss Samuel Johnson, Richard Kirwan, and Francis Galton in greater detail in Pickover, Clifford, *Strange Brains and Genius: The Secret Lives of Eccentric Scientists and Madmen* (New York: Quill, 1999).

12. Ibid.

13. Ferris, Timothy. *The Whole Shebang* (New York: Simon & Schuster, 1997), 312.

14. I also enjoy talking about "silence" in my books *The Paradox of God* (New York: Palgrave, 2001) and *The Stars of Heaven* (New York: Oxford University Press, 2001).

15. *Der Salon* is a collection of miscellaneous writings in four volumes published by Heinrich Heine in 1836–40.

16. Thomas Beecham is generally believed to be the finest British conductor of the early twentieth century.

Chapter 3. Gilgamesh, God, and the Language of Angels

1. Smith, Audrey. "Studies on Golden Hamsters During Cooling and Rewarming from Body Temperatures below 0 Degrees Centigrade," *Proceedings of the Royal Society of Biology, London Series B*, 147: 517, 1957.

2. Suda, Isamu, and Kito, Adachi C. "Histological Cryoprotection of Rat and Rabbit Brains," *Cryoletters* 5: 33, 1966.

3. Suda, Isamu and Kito, Adachi, C. "Bioelectric Discharges of Isolated Cat Brain after Revival from Years of Frozen Storage," *Brain Research*, 70 527–531, 1974.

4. Haan, E. A., and Bowen, D. M. "Protection of Neocortical Tissue Prisms from Freeze-Thaw Injury by Dimethyl-sulphoxide." *Journal of Neurochemistry*, 37: 243–246, 1981. Human brain tissue was excised during brain operations, requiring the removal of the cortex to allow access to deep tumors. For more information on freezing the brain, see "Declaration of a Leading Cryobiologist," www.cryonics.org/cryobiologist.html. (The identity of the author of this declaration has been withheld from this Web site.)

5. Storm, Howard. *My Descent into Death, and the Message of Love which Brought me Back* (East Sussex, United Kingdom: Clairview Books, 2000). To read an excerpt from Howard Storm's book, visit www.near-death.com/storm.html.

6. Atwater, P.M.H., "Is There a Hell? Surprising Observations About the Near-Death Experience," *Journal of Near-Death Studies*, 10(3), Spring 1992, www.cinemind.com/atwater/hell.html.

7. Litke, Sid. "Survey of Bible Doctrine: Angels, Satan, Demons," www.bible.org/page.asp?page_id=388.

8. Lewis, James. *Encyclopedia of Death and the Afterlife* (Farmington Hills, Missouri: Visible Ink, 1995).

9. Sandars, Nancy K. *The Epic of Gilgamesh: An English Version with an Introduction* (Penguin Classics) (New York: Penguin Books, 1972), 92.

10. While it is true that George Smith is frequently said to have torn off his clothes and run around naked upon realizing that *Gilgamesh* was an early version of Noah, many colleagues doubt it actually happened. Smith was quite celebrated for his discovery, and perhaps the Victorians would not have forgiven such a delirious streak in the buff.

11. The Staff of *Christian History Institute*. "George Smith and the New Noah," chi.gospelcom.net/DAILYF/2003/12/daily-12-03-2003.shtml.

12. Mitchell, Stephen. *Gilgamesh: A New English Version* (New York: Free Press, 2004). Excerpt at www.denverpost.com/Stories/0,1413,36%257E27%257E2574950,00.html.

13. Ibid.

14. George Smith's story is told in C.W. Ceram. *Gods, Graves and Scholars: The Story of Archeology*, 2nd ed. (New York: Knopf, 1967), Chapter 22.

15. Redwood, Daniel. "Interview with Elisabeth Kübler-Ross," 1995, www.drredwood.com/interviews/kubler-ross.shtml.

16. Rosen, Jonathan. "The Final Stage," *The New York Times Magazine*, Section 6, 14, December 26, 2004. Note that in the 1970s, Kübler-Ross became entangled in a scandal when a psychic at her California retreat center had sex with bereaved widows who firmly believed that they were making love to the spirits of their husbands. Kübler-Ross denounced the psychic but still believed in the existence of life after death.

17. van Lommel, Pim, Ruud van Wees, Vincent Meyers, and Ingrid Elfferich. "Near-Death Experience in Survivors of Cardiac Arrest: a Prospective Study in the Netherlands," *Lancet*, 358 (2001): 2039–45. See also Robert Todd Carroll. "The Skeptic's Dictionary–Near Death Experience," http://skepdic.com/nde.html.

18. Blackmore, Susan, and Tom Troscianko. "The Physiology of the Tunnel," *Near-Death Studies* 8 (1989): 15-28. See also Morse, J., Castillo, P., Venecia, D., Milstein, J., and Tyler, D. "Childhood Near-Death Experiences," *American Journal of Diseases of Children*, 140 (1986): 1110–1114. Susan Blackmore. "Near-Death Experiences: In or Out of the Body?" *Skeptical Inquirer*, 16 (Fall 1991): 34–45.

19. Rosen, Jonathan, op. cit. "The Final Stage."

20. Ibid.

21. Brooks, Michael. "Interview: Return of 'A Beautiful Mind,'" *New Scientist*, 2478: 46, December 18, 2004.

22. A wonderful introduction to the life of William James comes from: Frank Pajares. "William James: Our Father Who Begat Us," *Educational Psychology: A Century of Contributions*, Barry J. Zimmerman and Dale H. Schunk, editors (Mahwah: New Jersey: Earlbaum, 2002), 41–64. On the Web: www.emory.edu/EDU CATION/mfp/PajaresJames.PDF.

23. Brooks, Michael, op. cit. "Interview: Return of 'A Beautiful Mind.'"

24. Ralph Barton Perry *The Thought and Character of William James* (Nashville, Tennessee: Vanderbilt University Press; reprint edition, 1996).

25. James, William. *The Varieties of Religious Experience: A Study in Human Nature* (New York: Random House, 1902 and 1929), 508–509.

26. James, William, "Human Immortality," in *The Works of William James: Essays in Religion and Morality*. Introduction by John J. McDermott (Cambridge, Massachusetts: Harvard University Press, 1982).

27. Acton, Alfred. *The Letters and Memorials of Emanuel Swedenborg* (Bryn Athyn, Pennyslvania: Swedenborg Scientific Association, 1948), II, p. 696. See also: Jane K. Williams-Hogan, "Swedenborg: A Biography," in *Swedenborg and His Influence*, ed. Erland J.

Brock (Bryn Athyn, Pennsylvania: The Academy of the New Church, 1988). www.glencairnmuseum.org/jkwh.html.

28. Tyson, Donald. *Scrying for Beginners* (St. Paul, Minnesota: Llewllyn), 253.

29. Acton, Alfred, op. cit. *The Letters and Memorials of Emanuel Swedenborg.*

30. Swedenborg, Emanuel, www.swedenborg.net/. See also, Emanuel Swedenborg, and George F. Dole (translator) *Heaven and Hell* (West Chester, Pennsylvania: Swedenborg Foundation, 1990); The Editors of The World Almanac, *The World Almanac Book of the Strange* (New York: New American Library, 1977), 255.

31. Swedenborg, Emanuel, Leonard Fox (editor), Donald L. Rose (translator), and Jonathan Rose (translator). *Conversations With Angels: What Swedenborg Heard in Heaven*, (West Chester, Pennsylvania: Swedenborg Foundation, 1996). Robert H. Kirven, *Angels in Action: What Swedenborg Saw and Heard.* (West Chester, Pennsylvania: Swedenborg Foundation, 1995).

32. See note 30.

33. Hort, G. M. *Dr. John Dee: Elizabethan Mystic and Astrologer* (Whitefish, Montana: Kessinger Publishing Company, 1997).

34. McKenna, Terence. *The Archaic Revival: Speculations on Psychedelic Mushrooms, the Amazon, Virtual Reality, UFOs, Evolution, Shamanism, and the Rebirth of the Goddess* (San Francisco: Harper, 1992).

35. Laycock, Donald C. and Stephen Skinner. *The Complete Enochian Dictionary: A Dictionary of the Angelic Language As Revealed to Dr. John Dee and Edward Kelley.* (Boston: Samuel Weiser, 1994). See also, "Alphabets Magical," www.geocities.com/SoHo/Lofts/2763/witchy/alphabets.html. (This font is freeware made by the Digital Type Foundry.)

36. Sandars, Nancy K., op. cit. *The Epic of Gilgamesh*, 34.

37. Ibid.

38. "Gilgamesh Tomb Believed Found," BBC News, April 29,

2003, http://news.bbc.co.uk/1/hi/sci/tech/2982891.stm. Note that the Sumerian font is by Guillaume Malingue, "Sumerian Font Page," http://gmalingue.free.fr/UrIII/UrIII/. Malingue had been studying Sumerian epigraphy and was unable to find useful Sumerian cuneiform fonts, so he decided to create his own font and provide it "freely to the Sumerian community"!

39. Mitchell, Stephen, op. cit. *Gilgamesh*, 62.

40. Ibid., 63

41. Le Guin, Ursula K., *The Dispossessed* (New York: Perennial 2003)

42. Wagner, Jon. "MonsterHunter Review of *Donovan's Brain*," http://monsterhunter.coldfusionvideo.com/Donovan's_Brain.html.

43. Kübler-Ross, Elisabeth in Daniel Redwood. "Interview with Elisabeth Kübler-Ross," 1995, www.drredwood.com/inter views/kubler-ross.shtml.

44. Hanson, Robin. "Fourteen Wild Ideas: Five of Which Are True!" http://hanson.gmu.edu/wildideas.html.

45. Kreps, Joel Ibrahim, writing in *Islamica*. Portions of this article are reprinted in "Muslims and Depression," *The New York Times*, Week in Review, Section 4, p. 12, March 27, 2005.

Chapter 4. *The Matrix*, Quantum Resurrection, and the Quest for Transcendence

1. Flannery-Dailey, Frances. "Robot Heavens and Robot Dreams: Ultimate Reality in A.I. and Other Recent Films," *Journal of Religion and Film*, 7(2), October 2003, www.unomaha.edu/jrf/Vol7No2/robotHeaven.htm.

2. Lovecraft, H. P. "The Whisperer in Darkness," written in 1930, published in *Weird Tales*, 18(1): 32–73, August 1931.

3. Egan, Greg. *Permutation City* (New York: Eos, 1995), 252.

4. Davis, Owen. "Palynology Definitions," www.geo.arizona.edu/palynology/ppalydef.html.

5. Hanson, Robin. "If the Uploads Come First: The Crack of a Future Dawn," http://hanson.gmu.edu/uploads.html. Article originally appeared in *Extropy* 6 (1994): 2.

6. Harris, Sam, *The End of Faith* (New York: W.W. Norton & Company, 2004), 41.

7. van Eeden, Frederik. "A Study of Dreams," published in the *Proceedings of the Society for Psychical Research*, Vol. 26, 1913.

8. Gaydos, Chuck. personal communication.

9. Pickover, Clifford. *Sex, Drugs, Einstein, and Elves* (Petaluma, California: Smart Publications, 2005).

10. Osborne, Lawrence. "Inward Bound: Stephen LaBerge Offers the Ultimate Dream Vacation, Teaching You to Control What Unfolds In Your Mind While You Sleep," *The New York Times Magazine*, Section 6: 36–39, July 18, 2004.

11. I discuss reality simulation in Pickover, Clifford, *The Möbius Strip: Dr. August Möbius's Famous Band in Mathematics, Games, Literature, Art, Technology, and Cosmology* (New York: Thunder's Mouth Press, 2006).

12. Rees, Martin. "Living in a Multiverse," in *The Far Future Universe*, edited by George Ellis (West Conshohocken, Pennsylvania, Templeton Press, 2002), 65–88. Also see Nick Bostrom, www.simulation-argument.com, and Rees, Martin. "Edge: In the Matrix," www.edge.org/3rd_culture/rees03/rees_p2.html.

13. Davies, Paul. "A Brief History of the Multiverse," *The New York Times*, Late Edition–Final, Section A , Page 13, Column 3, April 12, 2003, Saturday. Also see the Nick Bostrom Web page www.simulation-argument.com.

14. Egan, Greg, op. cit. *Permutation City*.

15. Doctorow, Cory. "Thought Experiments: When the Singularity Is More Than a Literary Device: An Interview with Futurist-Inventor Ray Kurzweil," *Asimov's Science Fiction*, http://asimovs.com/_issue_0506/thoughtexperiments.shtml.

16. Strout, Joe. "Mind Uploading Home Page," www.ibiblio.org/jstrout/uploading/MUHomePage.html. See also Byrne, J. H., et al. "Neural and Molecular Bases of Nonassociative and Associative Learning in Aplysia," *Annals of the New York Academy of Sciences*, 627 (1991): 124-49.

17. Kurzweil, Ray. "The Human Machine Merger: Are We Headed for the Matrix?" In *Taking the Red Pill: Science, Philosophy and Religion in The Matrix*, Glenn Yeffeth (editor) (Dallas, Texas: Benbella Books, 2003).

18. Ibid.

19. "Poem 410." Some scholars refer to poems by numbers in Thomas Johnson's *The Complete Poems of Emily Dickinson* (Boston, Massachusetts: Back Bay Books, 1976).

20. Miller, Mark. "Matrix Revelations: The Wachowski Brothers FAQ," *WIRED*, 11(11), November 2003, www.wired.com/wired/archive/11.11/matrix.html.

21. Ibid. *WIRED* puts one of the brothers high on the chameleonic index by stating that he is likely to be either: 1) taking female hormones to prepare for a sex change operation, or 2) is a cross-dresser, or 3) is a transexual. *WIRED* says that one source dismisses the sex change rumor, explaining that the brother in question is merely a cross-dresser, not a transsexual.

22. Zynda, Lyle. "Was Cypher Right?" In *Taking the Red Pill: Science, Philosophy and Religion in The Matrix*, Glenn Yeffeth (editor) (Dallas, Texas, Benbella Books, 2003).

23. Chalmers, David J. "The Matrix as Metaphysics," http://consc.net/papers/matrix.html.

24. Davidson, Keay. "Cosmic Computer—New Philosophy to Explain the Universe," *San Francisco Chronicle*, July 1, 2002.

25. Ibid.

26. Bostrom, Nick. "Are We Living in The Matrix? The Simulation Argument," In *Taking the Red Pill: Science, Philosophy and Religion in The Matrix*, Glenn Yeffeth (editor) (Dallas, Texas: Benbella Books, 2003).

27. Bostrom, Nick. "The Simulation Argument: Why the Probability that You Are Living in a Matrix Is Quite High," *Times Higher Education Supplement*, May 16, 2003, www.simulation-argument.com/matrix.html.

28. Ibid.

29. Bostrom, Nick, op. cit. "Are We Living in The Matrix?"

30. Davies, Paul, "Higher laws and the Mind-boggling Complexity of Life: The Sum of the Parts," *New Scientist*, 2489, March 5, 2005.

31. Bostrom, Nick "Are You Living in a Computer Simulation?" *Philosophical Quarterly* 53(211): 243–255, 2003. www.simulation-argument.com/simulation.html.

32. Ibid.

33. Krauss, Lawrence M., and Glenn D. Starkman. "Universal Limits on Computation," *Physical Review Letters*, submitted, http://arxiv.org/abs/astro-ph/0404510.

34. Chown, Marcus. "Random Reality," *New Scientist*, February 26, 2000, http://members.fortunecity.com/templarser/randreal.html.

35. Ibid.

36. Ibid.

37. "Process Physics," Wikipedia, http://en.wikipedia.org/wiki/Process_Physics.

38. Chown, Marcus, op. cit. "Random Reality."

39. "Process Physics," Wikipedia, op. cit.

40. Gent, Charles "Is Reality a Side Effect of Randomness?", *Flinders Journal*, 11(4), March 27–April 9, 2000.

41. Battersby, Stephen. "Quantum resurrection," sidebar in "The Final Unraveling of the Universe," *New Scientist*, 185(2485): 31–37, February 5, 2005.

42. Ibid.

43. Moravec, Hans (interview). "Robot Children of the Mind," in David Jay Brown's *Conversations on the Edge of the Apocalypse* (New York: Palgrave, 2005), 121–136.

44. Barrow, John. "Living in a Simulated Universe," in *Universe or Multiverse*, edited by B. J. Carr (New York: Cambridge University Press, 2005).

45. Sandberg, Anders. "The Singularity,"
www.aleph.se/Trans/Global/Singularity/.
46. Harris, Sam. *The End of Faith* (New York: Norton, 2004), 35.
47. Ibid., 41

Chapter 5. Jesus and the Future of Mind-Altering Drugs

1. Terence McKenna (1946–2000) was a writer and philosopher
who was frustrated by modern society's ban on hallucinogens,
which he felt opened doorways to other worlds and could be
used responsibly by explorers of these worlds. In *Food of the
Gods*, McKenna advanced the theory that the use of psilocybin
mushrooms by primitive humans pushed evolution to the development of modern humans. Small doses of psilocybin enhanced
visual acuity, and thus primates that lived in areas where the
mushrooms thrived, gained an evolutionary advantage in
hunting. McKenna also theorizes that psilocybin-induced synes-
thesia (blurring of the senses) led to the development of spoken
language by triggering pictures in another person's mind by
vocal sounds.

2. "Cannabis linked to Biblical Healing," *BBC News*, 2003,
http://news.bbc.co.uk/2/hi/health/2633187.stm.

3. Ibid.

4. Bednar, Tim. "Junk Scholarship, Jesus, Cannabis, Preaching,"
www.e-church.com/Blog-detail.asp?EntryID=8&BloggerID=1.
Also see Chris Bennett and Neil McQueen, *Sex, Drugs, Violence
and the Bible*, (Vancouver, Canada: Forbidden Fruit Publishing,
2001).

5. Fabbro, Franco. "Did Early Christians Use Hallucinogenic
Mushrooms? Archeological Evidence," 1996,
http://people.etnoteam.it/maiocchi/fabbro.htm.

6. Ruck, Carl. "Was There a Whiff of Cannabis about Jesus?"
The Sunday Times, January 12, 2003,
www.cannabis.net/articles/jesus-cannabis.html.

7. Bergman, Ingmar, quoted in John Berger, "Ev'ry Time We Say Goodbye," *Sight and Sound* (a British film journal, London, 1991).

8. Jameson, Frederic. *Signatures of the Visible* (New York: Routledge, 1990).

9. Luciano, Patrick. *Them or Us: Archetypal Interpretations of Fifties Alien Invasion Films* (Bloomington: Indiana University Press, 1987).

10. Corman, Roger. *How I Made a Hundred Movies in Hollywood and Never Lost a Dime* (New York: Da Capo Press, 1998), 38.

11. Ibid., 5.

12. "Rober Corman biography," New Concorde, www.newconcorde.com/roger_corman.htm.

13. Taqi, S. "The Drug Cinema," http://leda.lycaeum.org/?ID=16668.

14. The Editors. "Marijuana Research," *Scientific American*, p. 8, December 2004.

15. Bennett, Drake. "Dr. Ecstasy," *The New York Times Magazine*, Section 6: 30–37, January 30, 2005.

16. Szalavitz, Maia. "Give Us the Drugs," *New Scientist*, 19, January 29, 2005.

17. Lawton, Graham. "Cannabis: Too Much, Too Young?" *New Scientist*, 2492: 44, March 26, 2005.

18. Ibid.

19. Ibid.

20. "Dope Dilemma?" *New Scientist*, 186: 2495, April 16, 2005.

21. Lawton, Graham, op. cit. "Cannabis: Too Much, Too Young?"

22. Horgan, John. "The Electric Kool-Aid Clinical Trial," *New Scientist*, 185(2488): 36–39, February 26, 2005.

23. "Ecstasy May Trigger Gene-Linked Depression," *New Scientist*, 185(1490), March 12, 2005.

24. de Alverga, Alex Polari. *O Livro das Mirações* (*The Book of Visions*), (Rio de Janerio: Editora Record, 1984). An English

translation is now available: Alex Polari de Alverga, *Forest of Visions: Ayahuasca, Amazonian Spirituality, and the Santo Daime Tradition.* Edited and introduced by Stephen Larsen. (Rochester, Vermont: Park Street Press, 1999).

25. Horgan, John, op. cit. "The Electric Kool-Aid Clinical Trial."

26. Ibid.

27. Vastag, Brian. "Ibogaine Therapy: A 'Vast, Uncontrolled Experiment'," *Science*, 308(5720): 345–346, April 15, 2005.

28. Ibid.

29. Leonard, Andrew. "California Dreaming: A True Story of Computers, Drugs and Rock 'n' Roll," *The New York Times*, Metro Section, B17, May 7, 2005.

30. Ibid.

31. Friedman, Thomas. TV interview "In Depth: Thomas Friedman," C-Span2, Book TV, May 1–2, 2005.

32. T. G., personal communication, May 7, 2005.

33. The Editors, op. cit. "Marijuana Research."

34. Ibid.

35. Abrams, Donald, Carroll Child, Thomas Mitchell. "Marijuana, the AIDS Wasting Syndrome, and the US Government," *New England Journal of Medicine* 333(10): 670–671, September 7, 1995.

36. Ibid.

37. Corral, Valerie (interview). "Medical Freedom and Cannabis Consciousness," in David Jay Brown's *Conversations on the Edge of the Apocalypse* (New York: Palgrave, 2005), 233–241.

38. Wilson, Clare. "Miracle Weed," *New Scientist* 185(2485): 38–41, February 5, 2005.

39. Nicoll, Roger, and Bradley Alger. "The Brain's Own Marijuana," *Scientific American*, 69–75, December 2004.

40. Ibid.

41. Ibid.

42. Ibid.

43. Wilson, Clare, op. cit. "Miracle Weed."

44. Nicoll, Roger, and Bradley Alger, op. cit. "The Brain's Own Marijuana."

45. Wilson, Clare, op. cit. "Miracle Weed."

46. Philips, Helen, and Graham Lawton. "The Intoxication Instinct," *New Scientist*, 32–42, November 13, 2004.

47. Ibid.

48. Ibid.

49. Ibid.

50. Tart, Charles. "States of Consciousness and State-Specific Sciences," *Science*, 176: 1203–1210, 1972.

51. Cronkite, Walter. "Prisons Needlessly Overpopulated with Drug Offenders," *Centre Daily Times*, August 6, 2004, www.mapinc.org/tlcnews/v04/n1118/a03.html?116519.

52. Ibid.

53. Ibid.

54. Ibid.

55. Ibid.

56. Larsen, Dana. "Global Ganja Executions," January 5, 2005, www.cannabisculture.com/articles/4124.html.

57. Ibid.

58. "Rusty," personal communication.

59. Brinkley, Joel. "Stopping Illicit Drugs Is Still Uphill Battle, Report Shows," *The New York Times*, Section A5, March 8, 2005.

60. Ibid.

61. Alexander Shulgin in Drake Bennett's "Dr. Ecstasy," *The New York Times Magazine*, Section 6, 33–37, January 30, 2005.

62. J. Edgar Hoover, former Director of the Federal Bureau of Investigation, quoted in *The Western City Magazine*, May 1938.

63. This quotation comes from "Drug War Is Lost," *New Scientist*, 185(2490): 4, March 12, 2005. The Seattle-based King County Bar Association published a 146-page report in 2005 recommending that the state should control production and

distribution of psychoactive drugs such as marijuana, cocaine, and heroin.

64. Campbell, R. "You Are What They Ate: A Brief Survey of Entheogenic Foods and Their Possible Roles in Human Evolution," www.v72.org/entheogen_cambell.htm.

Chapter 6. Clockwork Butterflies and Eternity

1. Mitchell, Stephen. *Gilgamesh: A New English Version* (New York: Free Press, 2004), 3.

2. McLemee, Scott. "Gods and Monsters," *Newsday*, October 17, 2004, www.mclemee.com/id129.html.

3. Rilke, Rainer Maria. *The Notebooks of Malte Laurids Brigge* (New York: Vintage, 1990).

4. Rilke, Rainer Maria. "Duino Elegies: The First Elegy," in *Ahead of All Parting: The Selected Poetry and Prose of Rainer Maria Rilke*, edited and translated by Stephen Mitchell (New York: Modern Library, 1995).

5. Hawthorne, Sophia. *Journal of Sophia Hawthorne*, January 14, 1851, Berg Collection, New York Public Library.

6. "Red-Letter Sales," *Writer's Digest*, 85(5): 12, May 2005.

7. Hawthorne, Nathaniel. "The Artist of the Beautiful," in *The Complete Novels and Selected Tales of Nathaniel Hawthorne*, edited by Norman Holmes Pearson (New York: Modern Library, 1937), 1139–1156. Story originally published in 1846.

8. Ibid.

9. Ibid.

10. Ibid.

11. Pickover, Clifford. *Strange Brains and Genius: The Secret Lives of Eccentric Scientists and Madmen* (New York: Quill, 1999).

12. Ibid.

13. Jacobs, A. J. *The Know-It-All: One Man's Humble Quest to Become the Smartest Person in the World* (New York: Simon and Schuster, 2004), 191.

14. Cromie, William J. "The Brains behind Writer's Block: New Views of the Muse," Harvard News Office, *Harvard Gazette Archives*, January 29, 2004, www.news.harvard.edu/gazette/2004/01.29/01-creativity.html.

15. Ibid.

16. Various books discuss writers who have used drugs. These include Marcus Boon, *The Road of Excess: A History of Writers on Drugs* (Cambridge: Harvard University Press, 2002); and Sadie Plant, *Writing on Drugs* (New York: Picador, 2001).

17. Burroughs, William, *Junkie* (London: Penguin, 1977).

18. Boon, Marcus. *The Road to Excess* (Cambridge: Harvard University Press, 2002).

19. Daumal, René. *Essais et notes* (2 volumes) (Paris: Gallimard, 1972).

20. Ibid.

21. Davenport-Hines, Richard. "Do Artists Need Narcotics Even More than Ordinary People?" *The Independent*, November 30, 2003, http://enjoyment.independent.co.uk/books/features/story.jsp?story=469086.

22. Ibid.

23. Jünger, Ernst. *Heliopolis*, 1949.

24. Zavrel, B. John. "Ernst Jünger: A Miracle at 100 Years," *PROMETHEUS, Internet Bulletin for Art, Politics and Science*, reprinted at www.meaus.com/JUENGER.html.

25. Kaku, Michio. "Parallel Universes, the Matrix, and Superintelligence," published on KurzweilAI.net, June 26, 2003, www.kurzweilai.net/meme/frame.html?main=/articles/art0585.html. Dr. Michio Kaku–theoretical physicist, author, and professor–maintains a wonderful Web site at www.mkaku.org/. Ray Kurzweil–inventor, entrepreneur, author, and futurist–also maintains an awesome Web site, www.kurzweilai.net/. Visit both authors' Web sites to feed your head.

Chapter 7. Evolution, Ice Cream, and the Goddess of Chopped Liver

1. Thomas Jr., Robert. "The Queen of Chopped Liver," in *The Last Word*, edited by Marvin Siegel (New York: William Morrow & Co, 1997), 26–27.

2. Lax, Andy. "Chopped Liver," www.sallys-place.com/food/single-articles/chopped_liver.htm.

3. Dollenmayer, Bob. John "Liver-Eating" Johnston, www.findagrave.com/cgi-bin/fg.cgi?Page=gr&GRid=4943

4. Shermer, Michael. *Science Friction* (New York: Times Books, 2005). Note that in other industrialized countries, 80 percent or more of those surveyed typically accept evolution. Of the remaining 20 percent, most say they are not sure, and very few people reject the idea outright.

5. Shermer, Michael. "The Fossil Fallacy," *Scientific American*, 32, March 2005.

6. I discuss these topics further in Clifford Pickover, *The Science of Aliens* (New York: Basic Books, 1998). Note that interstellar dust contains organic compounds but does not include the essential building blocks of Earth life, such as amino acids, sugars, fatty acids, and the bases of nucleic acids. For more information, see Shapiro, R. and Feinberg, G. "Possible Forms of Life in Environments Very Different from the Earth," in *Extraterrestrials: Where are They?* B. Zuckermanand M. Hart, eds. (New York: Cambridge University Press, 1995).

7. Calvin, Melvin. *Chemical Evolution: Molecular Evolution Towards the Origin of Life on the Earth and Elsewhere* (New York: Oxford University Press, 1969).

8. de Duve, Christian. *Vital Dust: Life as a Cosmic Imperative* (New York: HarperCollins, 1995).

9. This virtual typewriter conclusion assumes approximately 10^{86} particles in the universe and 20 billion years for the age of the universe.

10. A mathematical analysis of this problem is available upon e-mail request. Although this mathematical problem is a metaphor for evolution by maintaining adaptive improvements, in reality some adaptive improvements will not necessarily increase life expectancy unless they come in the proper sequence. Applying this rationale to the monkey that types Genesis means that the monkey may need to type certain letters first before he can move on to others.

11. Ardrey, Robert. *African Genesis* (New York: Macmillan, 1961).

12. "Born in Violence," Book Review of *African Genesis*, December 15, 1961, www.time.com/time/archive/preview/ 0,10987,827115,00.html.

13. Ardrey, Robert, op. cit. *African Genesis*.

14. Pennock, Robert. "The Very Model of Evolution?" Letter to the Editor, *Discover*, 8, April 2005.

15. Eaton, Joe. "Origin of a Species," *Terrain* magazine, www.terrainmagazine.org/article.php?id=13071.

16. Halliburton, Richard, and G. A. E. Gall. "Disruptive Selection and Assortative Mating in *Tribolium castaneum*," *Evolution*, 35 (1981): 829-843. See also Joseph Boxhorn, "Observed Instances of Speciation," http://talkorigins.org/faqs/faq-speciation.html.

17. Byrne, Katharine and Richard Nichols. "*Culex pipiens* in London Underground Tunnels: Differentiation between Surface and Subterranean Populations," *Heredity* 82 (1999): 7-15. See also "New Species," The Talk.Origins Archive, www.talkorigins.org/indexcc/CB/CB910.html; Alan Burdick, "Insect From the Underground–London, England Underground Home to Different Species of Mosquitoes," *Natural History*, February 2001, www.findarticles.com/p/articles/mi_m1134/is_1_110/ai_70770157.

Caveat: The mosquito work is not without controversy in the sense that several researchers have suggested that *molestus* and *pipiens* are different "behavioral" forms of the same species. *Pipiens* and *molesuts* indeed behave as different species in England, but further south (for example in Southern France) they seem to interbreed. Dr. Nichols discusses the reasons in his paper published in *Heredity*. He also writes to me:

> Underground mosquitoes from many different parts of the world appear to share similar genotypes. This seems to mean that the underground form has spread large distances with very little interbreeding with aboveground *pipiens* forms. Note that in the USA, the *Culex* appear to be some sort of hybrid, and bite both birds and humans, making them a bridging vector that may transmit diseases from birds to humans. See Wayne J. Crans, http://www.rci.rutgers.edu/~insects/pip2.htm.

Dr. Nichols later wrote to me:

> I would say there is strong evidence that the underground mosquitoes are on their way to speciation, but there is continuing gene flow in some parts of the species range. Hence this is a good candidate for speciation caught in the act. . . . Not yet completely reproductively isolated but some way along the path.

18. Shermer, Michael, op. cit. 181.
19. Ibid., 183.
20. Holt, Jim. "Unintelligent Design," *New York Times Magazine*, Section 6, pp 15–16, February 20, 2005.
21. Ibid.
22. Finn, J. T. "Birth Control Pills Cause Early Abortions," updated April 23, 2005, www.prolife.com/ BIRTHCNT.html.

23. Grayling, A. C. "The Power to Choose a Baby's Gender," *New Scientist*, 186(2494): 17, April 9, 2005.

24. Much of the information about liver divination comes from my book *Dreaming the Future* (Amherst, New York: Prometheus, 2002). The Cicero quote is from Marcus Tullius Cicero and C. D. Yonge (translator), *The Nature of the Gods and on Divination* (Great Books in Philosophy) (Amherst, New York: Prometheus, 1997).

25. Gurney, O. R. "The Babylonians and Hittites," In Michael Loewe and Carmen Blacker, *Oracles and Divination*, (Boulder: Shambala, 1981), 148.

26. Ibid., 149.

27. Shakespeare, William, *Julius Caesar*, Act 1, Scene 2, 15–24.

28. Wilson, Damon *The Mammoth Book of Nostradamus and Other Prophets* (New York: Carroll & Graf, 1999), 20.

29. Shakespeare, William, *Love's Labour's Lost*. Act 4, Scene 3. The character is saying something like, "This is what love is all about. It makes a man see a mortal as a deity, a young woman (green goose) as a goddess."

30. "Food Facts and Trivia: Delicatessen," www.foodreference.com/html/fdelicatessen.html.

31. Jacoby, Marvin. *The New York Times Magazine* (Letters section), Section 6, p. 10, March 6, 2005.

32. Lawton, Graham. "Life's Greatest Invention," *New Scientist* 186(2494): 28, April 9, 2005. (A section that gives many different scientists' opinions.)

Chapter 8. The Whispers of History

1. Here are the solutions for the Nicaragua stamp list: 1) Basic addition formula. 2) Isaac Newton's law of universal gravitation. If the two masses m_1 and m_2 are separated by a distance, r, the force exerted by one mass on the other is F, and G is a gravitational constant of nature. 3) Einstein's formula for the conversion of

matter to energy. 4) John Napier's logarithm formula. This allows us to do multiplication and division simply by adding or subtracting the logarithms of numbers. 5) Pythagorean theorem relating the lengths of sides of a right triangle. 6) Bolzmann's equation for the behavior of gases. 7) Konstantin Tsiolkovskii's rocket equation, which gives the speed of a rocket as it burns the weight of its fuel. 8) de Broglie's wave equation, relating the mass, velocity, and wavelength of a wave-particle. h is Planck's constant. de Broglie postulated that the electron has wave properties, and that material particles have associated with them a wavelength. 9) Equation relating electricity and magnetism, derived from Maxwell's equations that describe electromagnetic waves including radio, radar, light, ultraviolet waves, heat radiation and X-rays. 10) Archimedes' lever formula.

2. Sartre, Jean-Paul. *Nausea* (New York: New Directions, 1964).

3. Malcolm X (with the assistance of Alex Haley). *The Autobiography of Malcolm X* (New York: Ballantine, 1990).

4. Brubacher, Abram. *The Volume Library* (New York: Educators Association, 1928).

5. Shermer, Michael. *Science Friction* (New York: Times Books, 2005).

6. Durant, Will, *The Lessons of History* (New York: MJF Books, 1997).

7. Ibid.

8. Pickover, Clifford. *The Möbius Strip* (New York: Thunder's Mouth Press, 2006).

9. Asimov, Isaac. *I. Asimov: A Memoir* (New York: Bantam, 1995).

10. Metzger, Richard. "The Crying of Liber 49: Jack Parsons, Antichrist Superstar," in Richard Metzger's *Book of Lies* (New York: The Disinformation Company, 2003), 198–201.

11. Gardner, Martin. *On the Wild Side* (Amherst, New York: Prometheus, 1992).

12. Shulman, Polly. "Rocket Man," *New York Times* (Book Review section), 20, March 6, 2005.

13. Metzger, Richard, op. cit. "The Crying of Liber 49: Jack Parsons, Antichrist Superstar," 200.
14. Butler, Brian. "Cameron: The Wormwood Star," in Richard Metzger's *Book of Lies* (New York: The Disinformation Company, 2003).
15. Ibid.
16. Jaspers, Karl. *The Origin and Goal of History* (London: Routledge & Kegan Paul Ltd, 1953).
17. Shermer, Michael, op. cit. *Science Friction.*
18. McLuhan, Marshall and Quentin Fiore. *War and Peace in the Global Village* (New York: Gingko Press, 2001).
19. Holt, Jim. "Doom Soon: A Philosophical Invitation to the Apocalypse," *Lingua Franca*, 7(8), October 1997. http://flatrock.org.nz/topics/environment/doom_soon.htm. Note that Holt's numbers were just estimates. It may be more accurate to say that today we are part of 6 billion of the 60 billion people who have ever walked the Earth.
20. Bostrom, Nick. "A Primer on the Doomsday Argument," www.anthropic-principle.com/primer1.html.
21. My colleagues enjoy debating the relevance of this train car example. For example, Graham Cleverley writes

> Whoever said anyone started the human race by tossing a fair coin? Even the Biblical creationists don't claim that. If there were a God, and God said, "I will only create x people," and you discovered that you were the x–1th, then of course Doomsday would be very close. But if God said, "I will create an infinite number of people," then no matter what number you were, Doomsday would be infinitely far away, and whatever number you were you would be very unusual.

22. A fragment from Mark Ganson's Doomsday simulator:

```
MakeRooms[] := Block[{r,c,f},
c = {"heads","tails"}; f = c[[Random[Integer,{1,2}]]];
If [f == "heads", r = Range[100];];
If [f == "tails", r = Range[10];];
Return[{f,r[[Random[Integer,{1,Length[r]}]]]}];
[];
```

```
counter = 0; numHeads = 0; numTails = 0; numSevens = 0;
numSevenHeads = 0; numSevenTails = 0;
m = {};
    While[True, m = makeRooms[ ]; counter++;
    If [m[[1]] == "heads", numHeads++;];
    If [m[[1]] == "tails", numTails++;];
    If [m[[2]] == 7, numSevens++;
    If [m[[1]] == "heads", numSevenHeads++;];
    If [m[[1]] == "tails", numSevenTails++;];
    [];
```

After letting the While[] loop run for a few minutes, the following results were obtained:

```
counter = 15,374,323;
numHeads = 7,686,471; numTails = 7,687,852; numSevens =
{845,408}
numSevenHeads = 76,781; numSevenTails = 768,627
numSevenHeads/(numSevenHeads+numSevenTails) =
76781/845408
numSevenTails/(numSevenHeads+numSevenTails)//N =
0.909179
```

The function makeRooms[ ] is the heart of the simulator. It creates either 10 rooms (train cars) or 100 rooms and returns a random room (either 1 through 10 or 1 through 100). It also returns either "heads" or "tails" depending on the result of a simulated coin flip. Three local variables are used in the function: r, c, and f. The r is the rooms list variable, which will be either $r = \{1,2,3. . . 9,10\}$ or $r = \{1,2,3. . . 98,99,100\}$, depending on the coin flip to come. The c variable is the virtual coin, $c = \{\text{"heads"}, \text{"tails"}\}$. The f variable represents the result of the flip, which will be $f = \text{"heads"}$ or $f = \text{"tails."}$ The return value of makeRooms[ ] will be a 2-element list in the form of $\{\text{"heads"}|\text{"tails"}, \text{randomInt}\}$. A sample return value might be $\{\text{"heads"}, 79\}$. Another might be $\{\text{"tails"}, 8\}$. Take special note of the Return[ ] statement, Return[$\{f,r [[\text{Random}[\text{Integer},\{1,\text{Length}[r]\}]]]\}$]. By using Length[$r$] we can generate a random number between 1 and 10 or between 1 and 100 depending on whether $r = \{1,2,3. . . 8,9,10\}$ or $r = \{1,2,3. . . 98,99,100\}$. In the While[ ] loop, we simply call makeRooms[ ] continuously until we tire of the affair and tabulate the results in a number of variables:

```
counter (* the total number of iterations *)
numHeads (* how many heads were tossed *)
numTails (* how many tails were tossed–roughly equal to
numHeads *)
{numSevens} (* how many 7s were produced *)
numSevenHeads (* if 7, how many {"heads",7} *)
numSevenTails (* if 7, how many {"tails",7} *)
(* probability *)
numSevenHeads/(numSevenHeads + numSevenTails)
(* show probability as a decimal *)
numSevenTails/(numSevenHeads + numSevenTails) // N
```

23. Bostrom, Nick. "A Primer on the Doomsday Argument," www.anthropic-principle.com/primer1.html. Note one possible counter-argument. If the universe is infinite and we can access

parallel universes, then *infinitely* many humans may exist, in which case some of the mathematical argument breaks down. Or perhaps we will learn to simulate an infinite number of realities, as discussed in Chapter 4. Perhaps population numbers will also dramatically decrease, which could affect our analyses. What other counter-arguments can you make?

One problem with the Doomsday Argument is that it's always been wrong in the past. The human population has been generally increasing since humans emerged on Earth, so any member of any generation could have used the Doomsday Argument to predict that the human race would end soon. We're here so that hasn't happened. How good can a predictive theory be when it has been wrong 100 percent of the time so far?

24. Kreis, Steven. "Lecture 21, The Utopian Socialists: Charles Fourier," www.historyguide.org/intellect/lecture21a.html.

25. Fourier, Charles. *Fourier: The Theory of the Four Movements* (Cambridge Texts in the History of Political Thought) Edited by Gareth Stedman Jones and Ian Patterson (New York: Cambridge University Press, 1996).

26. Klinkenborg, Veryln. "Johnson's Dictionary," *The New York Times*, Section 4, p. 13, April 17, 2005.

27. Dawkins, Richard, "The Atheist," Interview with Gordy Slack, Salon.Com, April 30, 2005, www.salon.com/news/feature/2005/04/30/dawkins/.

Conclusion

1. Calvino, Italo. *Le città invisibili*, translated by William Weaver as *Invisible Cities*. (London: Vintage, 1997).

2. Ibid.

3. Ibid.

4. Christina, Greta. "Comforting Thoughts about Death that Have Nothing to do with God," *Skeptical Inquirer*, March/April, 2005, p. 50.

5. Donne, John. "Meditation XVII," *Devotions Upon Emergent Occasions*, 1624.

6. Rucker, Rudy. *The Fourth Dimension: A Guided Tour of the Higher Universes*. (New York: Houghton Mifflin, 1985).

7. Chopra, Deepak. "Quantum Spirituality," in David Jay Brown, *Conversations at the Edge of the Apocalypse* (New York: Palgrave, 2005).

8. Ram Dass. "Here, Now and Tomorrow," in David Jay Brown, *Conversations at the Edge of the Apocalypse* (New York: Palgrave, 2005).

9. Rothstein, Mervyn. "Isaac Asimov, Whose Thoughts and Books Traveled the Universe, Is Dead at 72," *The New York Times*, April 7, 1992.

10. Gardner, Martin. "Science and the Unknowable," *Skeptical Inquirer*, 22(6): November/December 1998.

11. Einstein, Albert, quoted in Howard Eves's *Mathematical Circles Adieu* (Washington, DC: The Mathematical Association of America, 2002).

Cathedrals of the Mind

1. Barron, Bebe. Quote appears in *Incredibly Strange Music Volume II* (San Francisco, California: RE/Search, 1994), 194–200. See also www.researchpubs.com.

2. McGrath, Charles. "The Souped-Up, Knock-Out, Total Fiction Experience," *The New York Times*, Section 4, p. 16, April 17, 2005.

3. Wagner, Jon. "MonsterHunter Review of *Donovan's Brain*," http://monsterhunter.coldfusionvideo.com/Donovan's_Brain.html.

4. Baird, Abigail. "Sifting Myths for Truths About Our World," *Science*, 308(5726): 1261–1262, May 27, 2005. (A book review of *When They Severed Earth from Sky*.)

5. Ibid.

6. Albert Einstein described his summer home in the village of

Caputh in this manner. He lived here from 1929 to 1932. The
Nazis came to power in 1933.

7. Hanson, Robin. "Fourteen Wild Ideas: Five of Which Are
True!" 2001, http://hanson.gmu.edu/wildideas.html.

8. Kilts, Clinton, in "Mysteries of the Mind," *US News & World
Report*, February 28, 2005.

9. Doctorow, Cory, op. cit. "Thought Experiments: When the
Singularity Is More Than a Literary Device: An Interview with
Futurist-Inventor Ray Kurzweil."

10. Ibid.

11 Churchland, Patricia. "Brains Wide Shut?" *New Scientist*, 2497:
46, April 30, 2005.

12. Minkel, J. R. "Wanted: The Next Einstein," *Popular Science*,
266(6): 81, June 2005, 81. (Describes the work of neuroscientist
Karel Svoboda of Cold Spring Harbor Laboratory.)

13. Richard Dawkins. "The Atheist," Interview with Gordy Slack,
Salon.com, April 30, 2005, www.salon.com/news/
feature/2005/04/30/dawkins/.

14. Kaku, Michio. "Parallel Universes, the Matrix, and Superintelligence," published on KurzweilAI.net, June 26, 2003,
www.kurzweilai.net/meme/frame.html?main=/articles/art0585.ht
ml. Kaku is the author of numerous books, including *Parallel
Worlds: A Journey Through Creation, Higher Dimensions, and the
Future of the Cosmos*. Ray Kurzweil is the author of *The Singularity
Is Near: When Humans Transcend Biology*. Please buy their wonderful books at Amazon.com within ten minutes of reading
these words.

15. Ibid.

16. Ibid.

17. A majority of physicists accept the idea of multiple universes.
These particular statistics come from a poll of 72 leading physicists conducted by the American researcher David Raub in

1995 and published in the French periodical *Sciences et Avenir* in January 1998.

18. Henry Vaughan, *Silex Scintillans*, 1650.

19. Alexander Shulgin in Drake Bennett's "Dr. Ecstasy," *The New York Times Magazine*, Section 6, 33–37, January 30, 2005.

20. Kaku, Michio. "Parallel Universes, the Matrix, and Superintelligence," published on KurzweilAI.net, June 26, 2003, www.kurzweilai.net/meme/frame.html?main=/articles/art0585.html.

21. Lawton, Graham. "Life's Greatest Invention," *New Scientist* 186(2494): 28, April 9, 2005. (A section that gives many different scientists' opinions.)

22. Ibid.

23. "Food Facts and Trivia: Delicatessen," op. cit.

24. Nilsson, Dan. "Multi-eyed Jellyfish Casts New Light on Darwin's Puzzle," *New Scientist*, May 2005.

25. Dawkins, Richard. "The Atheist," Interview with Gordy Slack, Salon.com, April 30, 2005, www.salon.com/news/feature/2005/04/30/dawkins/.

26. Ibid.

27. Ibid.

28. June 2005 Harris poll of 1,000 American adults, reported in Claudia Wallis, "The Evolution War," *Time*, August 15, 2005, p. 28.

29. Ryerson, James. "Sidewalk Socrates," *The New York Times Magazine*, Section 6, 35, December 26, 2004.

30. "Pop Culture," from *Wikipedia, the Free Encyclopedia*, http://en.wikipedia.org/wiki/Pop_culture.

References

1. Raxworthy, Christopher J. "A Truly Bizarre Lizard," www.pbs.org/edens/madagascar/creature3.htm.

One vital characteristic of a highly creative person is that they have at least one other person in their life who doesn't think they are completely nuts.

—Vera John-Steiner, in Helen Phillips's
"Looking for Inspiration," *New Scientist*, October 29, 2005

I am still an agnostic, with pantheistic overtones. The sight of plants and children growing inclines me to define divinity as creative power. . . . I cannot reconcile the existence of consciousness with a deterministic and mechanistic philosophy. I am skeptical not only of theology but also of philosophy, science, history, and myself. I recognize supersensory possibilities but not supernatural powers.

—Will Durant in *Dual Autobiography*

Even the most fantastical ghost stories, including the old stories of religion, can serve a purpose. Whether they postulate superintelligent clouds of gas, insectoid aliens in hyperspace, a demiurge with multiple-personality disorder, or a loving God who for inscrutable reasons makes us suffer, well-told ghost stories can remind us of the unfathomable mystery at the heart of things. Our creation myths and eschatologies, our imaginings of ultimate beginnings and ends, can also help us discover our deepest fears and desires.

—John Horgan, "Between Science and Spirituality,"
The Chronicle Review, 2001

Both the scientist and the artist are trying to represent the reality beyond appearances. I believe that at the moment of creative insight, boundaries dissolve between disciplines and both artists and scientists search for new modes of aesthetics. That was certainly the case with Albert Einstein and Pablo Picasso. They were both trying to understand the true properties of space, and to reconcile them with how space is seen by different observers. Einstein discovered relativity and Picasso discovered cubism almost simultaneously.

—Arthur Miller, "One Culture," *New Scientist*, October 29, 2005

REFERENCES

❧ ☙

Almost half of the world's chameleon species live on the island of Madagascar. This chameleon community is not only the world's largest, it is also the world's most unique; with 59 different species existing nowhere outside of Madagascar.

    –Christopher J. Raxworthy, "A Truly Bizarre Lizard"[1]

If the sentence "This sentence is not true" is true, then it is not true, and if it is not true, then it is true.

This section focuses largely on book reference sources. The Notes section gives many additional references.

Acton, Alfred. *The Letters and Memorials of Emanuel Swedenborg* (Bryn Athyn, Pennyslvania: Swedenborg Scientific Association, 1948).

Ardrey, Robert. *African Genesis* (New York: Macmillan, 1961).

Asimov, Isaac, *I. Asimov: A Memoir* (New York: Bantam, 1995).

"Bipolar Link," *New Scientist*, 188(2526): 20, November 19, 2005. (Describes the 2005 work of Chang and Ketter that links bipolar disorder and creativity.)

Boon, Marcus. *The Road to Excess* (Cambridge, Massachusetts: Harvard University Press, 2002).

Brown, David. *Conversations at the Edge of the Apocalypse* (New York: Palgrave, 2005).

Brubacher, Abram. *The Volume Library* (New York: Educators Association, 1928).

Burroughs, William. *Junkie* (London: Penguin, 1977).

Butler, Brian. "Cameron: The Wormwood Star," in Richard Metzger's *Book of Lies* (New York: The Disinformation Company, 2003).

Calvin, Melvin. *Chemical Evolution: Molecular Evolution Towards the Origin of Life on the Earth and Elsewhere* (New York: Oxford University Press, 1969).

Calvino, Italo. *Le città invisibili,* translated by William Weaver as *Invisible Cities* (London: Vintage, 1997).

Capote, Truman. *In Cold Blood* (New York: Vintage, reprint edition, 1994).

Capote, Truman. *The Complete Stories of Truman Capote* (New York: Random House, 2004).

Corman, Roger. *How I Made a Hundred Movies in Hollywood and Never Lost a Dime* (New York: Da Capo Press, 1998).

Dainton, Barry. "Innocence Lost: Simulation Scenarios: Prospects and Consequences," 2002, www.simulation-argument.com/dainton.pdf.

De Bono, Edward. *Lateral Thinking: Creativity Step by Step* (New York: Perennial; reissue edition, 1973).

de Alverga, Alex Polari. *O Livro das Mirações (The Book of Visions)* (Rio de Janerio: Editora Record, 1984). An English translation is now available: Alex Polari de Alverga, *Forest of Visions: Ayahuasca, Amazonian Spirituality, and the Santo Daime Tradition.* Edited and introduced by Stephen Larsen. (Rochester, Vermont: Park Street Press, 1999).

de Duve, Christian. *Vital Dust: Life as a Cosmic Imperative* (New York: HarperCollins, 1995).

Durant, Will. *The Lessons of History* (New York: MJF Books, 1997).

Egan, Greg. *Permutation City* (New York, Eos, 1995).

Egan, Greg. "Burning the Motherhood Statements: Jeremy G Byrne and Jonathan Strahan Interview with Greg Egan," originally appeared *Eidolon* 11, January 1993, 18–30, http://eidolon.net/eidolon_magazine/issue_11/11_egan.htm.

Ferris, Timothy. *The Whole Shebang* (New York: Simon & Schuster, 1997).

Florida, Richard. *The Rise of the Creative Class* (New York: Perseus, 2002). (The paperback edition has a special preface.)

Elbert, Jerome. "Does the Soul Exist?" In Paul Kurtz's *Science and Religion* (Amherst, New York: Prometheus, 2003).

Fourier, Charles. *Fourier: The Theory of the Four Movements* (Cambridge Texts in the History of Political Thought), edited by Gareth Stedman Jones and Ian Patterson (New York: Cambridge University Press, 1996).

Fox, Douglas. "Brain box," *New Scientist,* 188(2524): 28–32, November 5, 2005. (Mentions the work of Olaf Sporns). See also, "Connectome." *New Scientist,* 188(2525): 62, November 12, 2005.

Gardner, Martin. *On the Wild Side* (Amherst, New York: Prometheus, 1992).

Hawthorne, Nathaniel. "The Artist of the Beautiful," In *The Complete Novels and Selected Tales of Nathaniel Hawthorne,* edited by Norman Holmes Pearson (New York: Modern Library, 1937), 1139–1156. Story originally published in 1846.

Herszenhorn, David. "Admission Test's Scoring Quirk Throws Balance Into Question," *The New York Times*, Section B1, p. 1, November 12, 2005. (Discusses how high school tests favor students who specialize. Mentions Stuyvesant High School.)

Hort, G. M. *Dr. John Dee: Elizabethan Mystic and Astrologer*, (Whitefish, Montana: Kessinger Publishing Company, 1997).

Jacobs, A. J. *The Know-It-All: One Man's Humble Quest to Become the Smartest Person in the World* (New York: Simon & Schuster, 2004).

James, William. *The Varieties of Religious Experience: A Study in Human Nature* (New York: Random House, 1902, 1929).

James, William. *Essays in Religion and Morality*. Introduction by John J. McDermott (Cambridge: Harvard University Press, 1982).

Jamison, Kay *Touched With Fire: Manic-Depressive Illness and the Artistic Temperament* (New York: The Free Press, 1993).

Jaspers, Karl. *The Origin and Goal of History* (London: Routledge & Kegan Paul Ltd, 1953).

Kaku, Michio. *Parallel Worlds: A Journey Through Creation, Higher Dimensions, and the Future of the Cosmos* (New York: Doubleday, 2004).

Keats, Jonathan. "John Koza has Built an Invention Machine," *Popular Science*, 268(5): 66–72, May 2006.

Kirby, Doug, Ken Smith, and Mike Wilkins. "Mega-Messiahs–Roadside America," www.roadsideamerica.com/rant/megamessiah.html.

Kirven, Robert H. *Angels in Action: What Swedenborg Saw and Heard*. (West Chester, Pennsylvania: Swedenborg Foundation, 1995).

Kotler, Stephen. "Extreme States," *Discover*, 26(7): 60–66, July 2005. Discusses the "benefits" of near-death experiences.

Kozinn, Allan. "Obituary: John Cage, 79, a Minimalist Enchanted With Sound, Dies," *The New York Times*, August 13, 1992, www.nytimes.com/learning/general/onthisday/bday/0905.html.

Krebs, Albin. "Obituary: Truman Capote is Dead at 59; Novelist of Style and Clarity," *The New York Times*, August 26, 1984, www.nytimes.com/learning/general/onthisday/bday/0930.html.

Kruglinski, Susan. "Big Blue to Build a Brain," *Discover*, 26(9): 9, September 2005.

Kurzweil, Ray. *The Singularity Is Near: When Humans Transcend Biology* (New York: Viking, 2005).

Laycock, Donald C. and Stephen Skinner. *The Complete Enochian Dictionary: A Dictionary of the Angelic Language As Revealed to Dr. John Dee and Edward Kelley* (Boston: Samuel Weiser, 1994).

Levitt, Steven, and Stephen Dubner. *Freakonomics: A Rogue Economist Explores the Hidden Side of Everything* (New York: William Morrow, 2005).

Lloyd, Seth. *Programming the Universe* (New York: Knopf, 2006).

Luciano, Patrick. *Them or Us: Archetypal Interpretations of Fifties Alien Invasion Films* (Bloomington: Indiana University Press, 1987).

Ludwig, Arnold. *The Price of Greatness: Resolving the Creativity and Madness Controversy* (New York: Guilford Press, 1995).

Malcolm X (with the assistance of Alex Haley). *The Autobiography of Malcolm X* (New York: Ballantine, 1990).

Markoff, John. *What the Dormouse Said: How the 60s Counterculture Shaped the Personal Computer* (New York: Viking, 2005).

McKenna, Terence. *The Archaic Revival: Speculations on Psychedelic Mushrooms, the Amazon, Virtual Reality, UFOs, Evolution, Shamanism, and the Rebirth of the Goddess* (San Francisco: Harper, 1992).

McLuhan, Marshall, and Quentin Fiore. *War and Peace in the Global Village* (New York: Gingko Press, 2001).

Metzger, Richard. "The Crying of Liber 49: Jack Parsons, Antichrist Superstar," in Richard Metzger's *Book of Lies* (New York: The Disinformation Company, 2003), 198–201.

Mitchell, Stephen. Gilgamesh: *A New English Version* (New York: Free Press, 2004).

Noble, Holcomb. "Elisabeth Kübler-Ross, Psychiatrist Who Revolutionized Care of Terminally Ill, Dies at 78," *The New York Times*, August 26, 2004, www.nytimes.com/2004/08/26/national/26kubler.html.

Pajares, Frank. "William James: Our Father Who Begat Us," *Educational Psychology: A Century of Contributions*, Barry J. Zimmerman and Dale H. Schunk, editors (Mahwah: New Jersey, Earlbaum, 2002), 41–64.

Pickover, Clifford. *Dreaming the Future* (Amherst, New York: Prometheus, 2002).

Pickover, Clifford. *The Möbius Strip: Dr. August Möbius's Famous Band in Mathematics, Games, Literature, Art, Technology, and Cosmology* (New York: Thunder Mouth Press, 2006).

Pickover, Clifford. *The Paradox of God and the Science of Omniscience* (New York: St. Martin's/Palgrave, 2001).

Pickover, Clifford. *Sex, Drugs Einstein and Elves* (Petaluma, California: Smart Publications, 2005).

Pickover, Clifford. *Strange Brains and Genius: The Secret Lives of Eccentric Scientists and Madmen* (New York: Quill, 1999).

Plant, Sadie. *Writing on Drugs* (New York: Picador, 2001).

"Pop Culture," *Wikipedia, the Free Encyclopedia*, http://en.wikipedia.org/wiki/Pop_culture.

Powell, Corey S. "Welcome to the Machine," *The New York Times Book Review*, Section 7, p. 19, April 2, 2006.

Redwood, Daniel. "Interview with Elisabeth Kübler-Ross," 1995, www.drredwood.com/interviews/kubler-ross.shtml.

Rees, Martin. "Living in a Multiverse," in *The Far Future Universe*, edited by George Ellis (West Conshohocken, Pennsylvania: Templeton Press, 2002).

Rilke, Rainer Maria. "Duino Elegies: The First Elegy," in *Ahead of All Parting: The Selected Poetry and Prose of Rainer Maria Rilke*, edited and translated by Stephen Mitchell (New York: Modern Library, 1995).

Sandars, Nancy K. *The Epic of Gilgamesh: An English Version With an Introduction* (Penguin Classics) (New York: Penguin Books, 1972).

Shermer, Michael. *Science Friction* (New York: Times Books, 2005).

Siegel, Marvin. *The Last Word: The New York Times Book of Obituaries and Farewells: A Celebration of Unusual Lives.* (New York: William Morrow & Company, 1997).

Storm, Howard. *My Descent into Death, and the Message of Love which Brought Me Back* (East Sussex, United Kingdom: Clairview Books, 2000).

Swedenborg, Emanuel, and George F. Dole (translator). *Heaven and Hell* (West Chester, Pennsylvania: Swedenborg Foundation, 1990).

Swedenborg, Emanuel, Leonard Fox (editor), Donald L. Rose (translator), and Jonathan Rose (translator). *Conversations With Angels: What Swedenborg Heard in Heaven* (West Chester, Pennsylvania: Swedenborg Foundation, 1996).

Thomas Jr., Robert. "The Queen of Chopped Liver," in *The Last Word*, edited by Marvin Siegel (New York: William Morrow & Co, 1997), 26–27.

Trivedi, Bijal. "Suspended Animation: Putting Life on Hold," *New Scientist*, 2535, January, 21, 2006, NewScientist.Com News Service, www.newscientist.com/channel/health/mg18925351.200.html. (Discusses pigs placed in suspended animation.)

Tyson, Donald. *Scrying for Beginners* (St. Paul, Minnesota: Llewllyn, 1997), 253.

Weeks, David and Kate Ward. *Eccentrics: the Scientific Investigation* (Stirling: Stirling University Press, 1988).

Williams, Caroline. "The 25 Hour Day," *New Scientist*, 189(2537): 34–37, February 4, 2006. (Discusses the brain's time perception, and the ideas of researchers Warren Meck, Catalin Buhusi, and Robert Levine.)

Williams-Hogan, Jane K. "Swedenborg: A Biography," in *Swedenborg and His Influence*, edited by Erland J. Brock (Bryn Athyn, Pennsylvania: The Academy of the New Church, 1988).

"William James Dies: Great Psychologist Brother of Novelist and Foremost American Philosopher Was 68 Years Old," *The New York Times*, August 27, 1910, www.nytimes.com/learning/general/onthisday/bday/0111.html.

Wilson, Damon. *The Mammoth Book of Nostradamus and Other Prophets* (New York: Carroll & Graf, 1999), 20.

Yeffeth, Glenn. *Taking the Red Pill: Science, Philosophy and Religion in The Matrix* (Dallas, Texas: Benbella Books, 2003).

❦ ❧

I would sooner live in a society governed by the first two thousand names in the Boston telephone directory than in a society governed by the two thousand faculty members of Harvard University.

—William F. Buckley, Jr., c. 1965

Everything evolves—or fails to evolve and is rendered obsolete. Ideas, technology, society rise and fall. Evolution occurs at every level above the quantum level, and this quantum level seems to be what allows it all to happen, this "flexibility of possibility."

—James Platt, personal communication

Gibraltar is the only place in Europe to have wild monkeys.

—A. J. Jacobs, *The Know-It-All*

A prism in the dark is just a hunk of glass.

—James Platt, personal communication

The only way for a person to have anything approaching consciousness after death. . . would be, while the person is alive, to learn to identify so profoundly with something other than his or her own ego so that when the self dies, the identification goes on. . . . The only way out [of death] would be to get out while you're here. I don't think you can get out after you're dead.

—Douglas Rushkoff, in David Jay Brown's *Conversations at the Edge of the Apocalypse*

INDEX
❧ ❦

ABOUT THE AUTHOR
❧ ❧

Clifford A. Pickover received his PhD from Yale University's Department of Molecular Biophysics and Biochemistry. He graduated first in his class from Franklin and Marshall College, after completing the four-year undergraduate program in three years. His many books have been translated into Italian, French, Greek, German, Japanese, Chinese, Korean, Portuguese, Spanish, Turkish, Serbian, Romanian, and Polish.

One of the most prolific and eclectic authors of our time, Pickover is author of the popular books: *The Möbius Strip* (Thunder's Mouth Press, 2006), *Sex, Drugs, Einstein and Elves* (Smart Publications, 2005), *A Passion for Mathematics* (Wiley, 2004), *Calculus and Pizza* (Wiley, 2003), *The Paradox of God and the Science of Omniscience* (Palgrave/St. Martin's Press, 2002), *The Stars of Heaven* (Oxford University Press, 2001), *The Zen of Magic Squares, Circles, and Stars* (Princeton University Press, 2001), *Dreaming the Future* (Prometheus, 2001), *Wonders of Numbers* (Oxford University Press, 2000), *The Girl Who Gave Birth to Rabbits* (Prometheus, 2000), *Surfing Through Hyperspace* (Oxford University Press, 1999), *The Science of Aliens* (Basic Books, 1998), *Time: A Traveler's Guide* (Oxford University Press, 1998), *Strange Brains and Genius: The Secret Lives of Eccentric Scientists and Madmen* (Plenum, 1998), *The Alien IQ Test* (Basic Books, 1997), *The Loom of God* (Plenum, 1997), *Black Holes: A Traveler's Guide* (Wiley, 1996), and *Keys to Infinity* (Wiley, 1995). He is also author of numerous other highly-acclaimed books including *Chaos in Wonderland: Visual Adventures in a Fractal World* (1994), *Mazes for the Mind: Computers and the Unexpected* (1992), *Computers and the Imagination* (1991), and *Computers, Pattern, Chaos, and Beauty* (1990), all published by St. Martin's Press—as well as the author of over 200 articles concerning topics in science, art, and mathematics. He is also coauthor, with Piers Anthony, of *Spider Legs*, a science-fiction novel once listed as Barnes and Noble's second best-selling science-fiction title. Pickover is currently an associate editor for the scientific journal *Computers and Graphics* and is an editorial board member for *Odyssey, Leonardo,* and *YLEM*.

Editor of the books *Chaos and Fractals: A Computer Graphical Journey* (Elsevier, 1998), *The Pattern Book: Fractals, Art, and Nature* (World Scientific, 1995), *Visions of the Future: Art, Technology, and Computing in the Next Century* (St. Martin's Press, 1993), *Future Health* (St. Martin's Press, 1995), *Fractal Horizons* (St. Martin's Press, 1996), and *Visualizing Biological Information* (World Scientific, 1995), and coeditor of the books *Spiral Symmetry* (World Scientific, 1992) and *Frontiers in Scientific Visualization* (Wiley, 1994), Dr. Pickover's primary interest is finding new ways to continually expand creativity by melding art, science, mathematics, and other seemingly disparate areas of human endeavor.

The *Los Angeles Times* has proclaimed, "Pickover has published nearly a book a year in which he stretches the limits of computers, art and thought." Pickover received first prize in the Institute of Physics "Beauty of Physics Photographic Competition." His computer graphics have been featured on the cover of many popular magazines, and his research has recently received considerable attention by the press—including CNN's "Science and Technology Week," The Discovery Channel, *Science News, The Washington Post, Wired,* and *The Christian Science Monitor*—and also in international exhibitions and museums. *OMNI* magazine described him as "Van Leeuwenhoek's twentieth century equivalent." *Scientific American* several times featured his graphic work, calling it "strange and beautiful, stunningly realistic." *WIRED* magazine wrote, "Bucky Fuller thought big, Arthur C. Clarke thinks big, but Cliff Pickover outdoes them both." Pickover holds over thirty US patents, mostly concerned with novel features for computers.

For many years, Dr. Pickover was the lead columnist for *Discover* magazine's Brain-Boggler column, and he currently writes the Brain-Strain column for *Odyssey*. His calendar and card sets, *Mind-Bending Visual Puzzles*, have been among his most popular creations.

Dr. Pickover's hobbies include the practice of Ch'ang-Shih Tai-Chi Ch'uan, Shaolin Kung Fu, and piano playing. He owns a 110-gallon aquarium filled with Lima shovelnose catfishes. These bizarre creatures resemble sharks with ultra-tiny, alien eyes. He advises readers to maintain a shovelnose tank in order to foster a sense of mystery in their lives. Look into the fish's eudaemonic eyes, dream of Elysian Fields, and soar.

Future books written in the spirit of *A Beginner's Guide to Immortality* may include topics in the following list. The topics receiving the most reader feedback will be selected: 1) Matrioshka Brains—megascale superintelligent thought machines, 2) quantum theory and the Carolingian Renaissance, 3) pareidolia and Marian apparitions, 4) Gram-Schmidt orthonormalization and the Dalai Lama, 5) the Gracchi brothers and the Phyllodocida, 6) the afterlife of Jesus—Oliver Messiaen and *Turangalila*, 7) the loneliness of factorion 40,585, 8) the medical mystery of Viking hero Egil Skallagrimsson—a poet at three and a killer at age seven, 9) aposiopesis and asyndeton in literature and life, 10) the history of calipee—the glutinous flesh next to a turtle's lower shell, and 11) the billion-dollar connectome of Olaf Sporns.

Visit Dr. Pickover's Web site, www.pickover.com, which has received over a million visits. He can be reached at this Web page or at P.O. Box 549, Millwood, New York 10546-0549 USA.

CPSIA information can be obtained at www.ICGtesting.com
Printed in the USA
LVOW10s1039240914

405547LV00001B/17/P